Flowers
of the
Canyon
Country

Flowers
of the
Canyon
Country

by
Stanley L. Welsh

Photography by
Bill Ratcliffe

University of Utah Press
Salt Lake City

Library of Congress Cataloging in Publication Data

Welsh, Stanley L.
 Flowers of the Canyon Country

 Includes index.
 1. Wild Flowers — Southwest, New — Identification.
I. Ratcliffe, Bill. II. Title.
 ISBN 0-87480-486-8

First edition 1971, Brigham Young University Press
Second edition 1977, revised and enlarged, Brigham Young University Press
University of Utah Press edition © 1986 Stanley L. Welsh
All rights reserved
Printed in Singapore

Paper cover photograph: Eaton beardtongue (*Penstemon eatonii* Gray), whose long tubular flowers must be pollinated by hummingbirds and other insects with long mouth parts (see pp. 36-37)

Scenic photographs
Page ii: Coral Pink Sand Dunes, northwest of Kanab, Kane County, Utah
Page vi: Sandblasted ponderosa pine, Coral Pink Sand Dunes, Kane County, Utah
Page xi: Sunrise on the Goblin Choir, Goblin Valley, northwest of Hanksville, in Emery County, Utah
Page xii: Angel Arch, stripped with desert varnish, Salt Canyon, Needles Section, Canyonlands National Park, Utah
Pages xiv-xv: West from Island-in-the-Sky, overlooking the canyons of the Green River, Canyonlands National Park, Utah
Page xvi: Castle Arch, Horse Canyon, Needles Section, Canyonlands National Park, Utah
Page xvii: Monument Valley, Navajo Indian Reservation, Arizona
Page xviii: Horsehoof Arch, Needles Section, Canyonlands National Park, Utah
Page 84:   Colorado River overview, Dead Horse Point State Park, Utah

Contents

Foreword

Wild flowers growing in the canyon lands of the western United States often seem paradoxical — blossoming oases in the midst of arid, craggy terrain. The canyon lands — located in the Four Corners region where the states of Arizona, Colorado, New Mexico, and Utah meet — are a harsh area, extreme in temperature, low in moisture, and rugged in configuration. Shaped by the timeless influence of wind and water, the area is well known for massive landforms and unimpeded vistas. But wild flowers contribute much to the beauty of this area as well, as this book lovingly attests.

Flowers of the Canyon Country is the result of the concentrated effort of two men who have been long interested in the wealth of beauty found in the West. Although they have collaborated previously on brief articles and features, the first edition of this book in 1971 was the first time they had united their talents to provide a book not only informative, scholarly, and artistic, but also indicative of the interest in the vast, largely unknown region of the Southwest.

The photography of Bill Ratcliffe is the work of an experienced professional. He spent two years work-ing on the Walt Disney production *Perri*, a wildlife feature photographed in the Uinta and Teton mountains. Mr. Ratcliffe's work has appeared in *Audubon* magazine and calendars, *Reader's Digest*, *Arizona Highways*, *National Wildlife*, *Pacific Discovery*, *American Forests*, *National Geographic*, and in such book publications as the *Encyclopaedia Britannica*, *Thorndike-Barnhardt Dictionary*, *Living Plants of the World* (Random House), *The American West* (Random House), *America the Beautiful* (Reader's Digest Books), *The Life of the Forest* (McGraw-Hill), and numerous textbooks.

He recently completed an assignment for *Time-Life* books on flowers. Some of his color prints appear in the Smithsonian Institution; the Monterey Peninsula Museum of Art, Carmel, California; and the Merritt Island National Wildlife Refuge, a display in the Kennedy Space Center. Bill Ratcliffe has tangibly demonstrated his interest in nature through his desire to share his appreciation for her intricacies.

Stanley L. Welsh, the author of the textual matter, started collecting plants in the Uinta Basin in 1953. From his extensive collections between 1955 and

1957 at the Dinosaur National Monument, he was able to complete a Master of Science degree at Brigham Young University. He has collected botanical information and specimens throughout the Great Plains, from the Texas Panhandle to Canada. His herbarium includes collections from all contiguous states west of the Mississippi and from the southeastern United States. He also has collected exhaustively in western Canada and Alaska.

Dr. Welsh completed his Ph.D. at Iowa State University in 1960. He helped identify plant remains from the Wetherill Mesa Project, Mesa Verde National Park, from 1961 to 1964, and he has written more than thirty scientific papers on plants of Utah national parks and monuments and on the flora of Alaska and the Yukon country. Presently a professor of botany at Brigham Young University, Dr. Welsh has collected plants in all counties of Utah, with special attention given to the canyon area and the three neighboring states of Colorado, New Mexico, and Arizona. He has also specialized in flora of the mountains of western North America north of Mexico. He has authored *Utah Plants: Tracheophyta*

(with Glen Moore); *Water, Stone, Sky: A Pictorial Essay on Lake Powell* (with Catherine Ann Toft); and *Anderson's Flora of Alaska and Adjacent Parts of Canada* (all published by Brigham Young University Press).

The scope of this book demonstrates the professional skill of these two men; the care indicates their scholarly competence. I welcome this revised and expanded edition of their 1971 volume, as well as their companion volume *Flowers of the Mountain Country* (Brigham Young University Press, 1975). Hopefully they will continue their joint efforts to illustrate and explain nature's treasures for an interested audience.

Glen Moore, Ph.D.

Using the Book

I have used technical terms sparingly, but they cannot be eliminated completely. Most people are familiar with the basic terminology of flowers, but for those who might wish to review the flower parts a discussion will be helpful.

Ordinarily flowers consist of four series of parts, arranged in ascending order and attached to the tip of the flower stalk. The outermost whorl consists of sepals; or if they are joined to each other or are discussed collectively, they are called the calyx. The calyx is usually green with leaflike parts and encloses the other floral structures in bud. The petals are borne interior to the sepals. They are usually brightly colored and have a different texture than the sepals. In some flowers the petals are joined into a tube which encloses the stamens and pistils. That tube, the corolla tube, occurs in a great number of shapes and sizes. The number of petals in a united corolla can be determined by counting the number of lobes at the tube apex or limb. In some flowers with tubular corollas the lobes flare, forming a definite limb of the corolla. Calyx and corolla together are known as a perianth, and in those flowers where only one whorl is present in the perianth that whorl is referred to arbitrarily as a calyx. Interior to the petals, and sometimes attached to them, are the stamens, which consist of elongate filaments and terminal anthers in which pollen is borne. The pistil is situated in the center of the flower. It consists of three basic parts: an apical, pollen-receptive stigma; a style; and an ovary. The ovary ripens to form the fruit.

Every aspect of a plant is subject to much variation: number of the various flower parts, their shape, size, and arrangement. The leaves, stems, roots, and other vegetative parts change also. Types of flower clusters, or inflorescence types, vary from one plant group to another. Some of the terms which describe the variability of each of these types of diversity are used in the treatment. Many of the terms are defined in the individual treatments, and it is not felt necessary or desirable to give a comprehensive review or glossary of the terms which describe all of the plant parts and variations.

It is not necessary to know any technical language to appreciate the beauty of the photographs or to identify common plants with the aid of this book. An interpretation of all of these complexities is not the role of this book, and it is not necessary for the reader to understand either the complex nature of these features or the technical language used to describe them. One can appreciate flowers for the charm of their shape and texture and for the emotion which he feels as a result of viewing the totality of light and shadow playing on a spectrum of color in a natural setting. This book is dedicated not just to presenting information on wild flowers but also to a description of a feeling for them. The more one knows of them the greater is the appreciation. Beauty is not intrinsic in the flowers themselves but lies in the imagination, the appreciation, and the understanding of the beholder.

Plants with similar flower color have been placed together, and one can identify species simply by comparing an unknown plant with those of similar flower color in the book. However, the reader should be aware that pigmentation in flowers is often very subtle; as light quality changes from dawn to dusk, color also changes. A particular color in midday light appears to be different in morning or evening light. As a result, flowers appear to be one color in shade and another in full sunlight. For example, white flowers take on a bluish cast in shade and a pinkish cast in full sunlight. Thus some variation in color must be allowed for when comparisons are being made.

Spelling of terms and common and scientific names is based upon that used in previous works by Stanley L. Welsh, *Utah Plants* and *Anderson's Flora of Alaska and Adjacent Parts of Canada*. Within each major color group, individual flowers are arranged alphabetically by scientific name.

*Remoteness, aridity, and
a tendency to penuriousness
are reasons for the wild charm and
beauty of the lands in and adjacent
to the canyons of the Colorado.*

The Canyon Country

Wild, natural beauty best describes the canyon country of the southwestern United States, particularly that area known as the Four Corners region of Arizona, Colorado, New Mexico, and Utah. This book will illustrate and discuss one of the prime sources of this beauty — the common native and introduced flowering plants of the canyon country. The Four Corners region, with its majestic scenery, deep canyons, magnificent colors, and immense vistas, reflects the vast periods of time and the surging forces of the wind and water that have created it.

The land is harsh, owing to extremes of temperature, low general precipitation, and natural ruggedness. Primarily the land is considered arid; however, the landforms vary in elevation from about 3,000 feet to about 12,000 feet. At the higher elevations the snows slowly melt during cool springs and summers, feeding the streams that flow to the Colorado or to some of its major tributaries. Where water is abundant, a verdure develops in great contrast to the nearby arid lands. Cottonwoods or poplars, willows, and other native and introduced shrubs and trees

grow along the streams. Abounding in wet spots, because of their large daily water requirements, are cattails, rushes, and other plants. However, the overall aspect of the land is controlled by the sparse cover of desert shrub vegetation and its predominating grays.

The massive sandstones forming the canyon cliffs are not uniformly sandy. Here and there are layers of finer-particled rock, the shales or mudstones or siltstones. These impervious layers block the downward flowing water which otherwise percolates readily through the sandstone, forcing it along the top of the shaly or silty layer until that stratum outcrops along a cliff face, there forming seeps and springs which allow for the growth of the peculiar vegetative hanging gardens. In other regions such spring and seep vegetation might not cause much notice, but in arid lands any spot of greenery is important and the hanging gardens form a unique element of the canyon country vegetation. Representative plants of North America — such as the cave primrose, sheathed death-camas, bracken, and maidenhair fern — are elements common to the

northern mountain region. Other plants have affinities with the vegetation of the southern and southeastern portions of the United States.

Aridity in the canyon country is a function of elevation, and vegetation is elevationally stratified in the region. At the lower elevations, in the bottoms of canyons in southern Utah and northern Arizona, shrubs dominate the hot desert. Between roughly 3,500 feet to 5,000 feet there is a cool desert shrubland of shadscale, saltbrush, and other species whose aspect is gray green in color. Interfingering with the cool desert shrubs and extending upwards to about 6,000- or 7,000-feet elevation is a pigmy-conifer forest of juniper and pinyon pine. The juniper and pinyon woodland is replaced on north-facing slopes by sagebrush and mountain brush species such as serviceberry, mountain mahogany, and shrubby species of oak. In other regions there is a zone of ponderosa pine above the pinyon and juniper woodland. This is in the position of the mountain brush zone or intermingling with it. At elevations above 6,500 to 7,500 feet the mountain brush gives way to aspen and spruce and fir woodland. Douglas-fir and aspen also occur at elevations below 6,500

feet on some steeply north-facing, moist slopes.

In the desert shrublands which occupy much of the lower regions of the canyon country, the land frequently has a barren appearance, as though there were no vegetation at all. A closer look will demonstrate an abundance of plants, some of surprising beauty. It is probable that the canyons, mesas, and mountains in the included area contain many more than one thousand species of native plants, and there are many additional introduced ones. Thus, an important part of understanding the totality of the beauty of the canyon country is in its flowers, and it is to these that this book is dedicated.

Remoteness, aridity, and a tendency to penuriousness are reasons for the wild charm and beauty of the lands in and adjacent to the canyons of the Colorado. Richness in beauty, charm, intrigue, and history have led to the establishment of a series of national parks and monuments dedicated to both preservation and exhibition of the beauties that combine to make an awesome display. Aridity and penuriousness of the land are tied one to the other, but the people of this generation owe the continuance of that land in an

undeveloped, unindustrialized state to those forces. The land has been considered to be good for nothing but grazing, and for the gentle tramp of the feet of a hardy few who would penetrate the region. Now, the situation is changing. Wealth in minerals within the canyon country has attracted the attention of a world demanding them, and that great sparingly inhabited canyon lands region is about to be impressed by all of the forces of an industrial society. Highways, rights-of-way, mines, industrial plants, towns, and semipermanent populations of people will modify the face of this magnificent land. The wild lands will shrink by an amount proportional to that used for these other purposes. The haze of a summer afternoon will be thickened by the accumulation of dust and other contaminants which accompany an increase in activity in this land of great vistas. Total quantity of vegetation will be lessened, and that remaining will be modified in order to reclaim areas cleared for the industrialization. Hopefully, the wildflowers of the canyon country will not only survive this change, but will provide both interest and enjoyment to all future generations.

The Flowers

Brightening the grays of the sparse desert cover, shaping the brilliant hanging gardens of the cliffs, and weaving a tapestry of color throughout the Southwest, the flowers of the canyon country — the primrose, the death-camas, the paintbrush, the four-o'clock, and others — reflect the wealth of nature. The flowers of the canyon country are beautiful and varied. One hundred species of flowering plants are illustrated on the following pages, including information on their distributional range, habitat, and flowering time. In most cases the approximate sizes of each plant and its flowers are given. There are some flowers, however, that are so well known to the general public that their size is not discussed.

Many plants are, in a sense, multipurpose. In such cases the common plant uses have been discussed; for instance, the Pueblo Indians of the eleventh and twelfth centuries used the yucca for everything from food to sandals. Some other plants, the jimsonweed, for example, have been used in religious ceremonies for their hallucinogenic effects — an overdosage could have been fatal. There are many medicinal plants, many poisonous plants, and some food plants. Where applicable, these qualities are indicated.

The flowers of the canyon country have an interesting and colorful history.

1

The sand-verbenas (here *Abronia fragrans* Nutt. ex Hook.) are attractive plants with white or, in some species, pink flowers in open, globose clusters. The fragrant perianths are from ¾ to 1¼ inches long and open wide in early evening. They remain open throughout the night, and their sweet odor scents the air. In mid-morning, the perianth of each flower closes, but the attractive umbels continue to brighten the sandy landscape through-out the day. They are clothed with hairs and with glands which secrete a sticky resinous substance. Sand and other soil particles, and even small insects, adhere to this secretion, often obscuring the green color of the herbage.

Small-leaved Pussytoes

Antennaria parvifolia Nutt.

The "flowers" of pussytoes are, like those of other composites, in reality clusters of flowers called heads which are surrounded by bracts. Unlike most other composites, however, the bracts of the heads in pussytoes are more showy than the flowers they enclose. The stems of small-leaved pussytoes are commonly less than 2 inches tall and the heads are about 1/3 of an inch long; the flowers within the heads are very tiny. Each plant will produce only male or female flowers, not both. The male flowers are hardly necessary, however, because the plants reproduce very well vegetatively and since seed is produced without the necessity of pollination having taken place. Plants which produce seed vegetatively are said to be "apomictic." This species grows in the juniper-pinyon, ponderosa pine, and mountain brush zones. It occurs from British Columbia and Manitoba southward to Arizona and New Mexico. The genus *Antennaria* consists of only 25 to 30 species and is mainly distributed in western North America. However, there are some that are circumpolar and others confined to South America.

Flowering time: May to August   Composite family

Prickly-poppy

Argemone corymbosa Greene

The generic name *Argemone* is derived from the Greek word *argema*, or cataract, a disorder of the eye which it was said to cure. The plants contain medicinally active ingredients, alkaloids, and have been used to treat various human disorders. The primary uses have been as a purgative, as a substitute for ipecac, and for treating disorders of the eyes. There are at least 23 species known from North America, some additional ones from South America, and one in Hawaii. Some of the species have been introduced into the tropical and subtropical regions of the world, where they persist. Plants of *Argemone corymbosa* are thistlelike. They grow to form clumps from about 8 inches to 3 feet tall. The size is dependent on moisture and soil conditions. The area of distribution is from southern Utah to Arizona and California. Flowers are 1½ to 4 inches wide. As in other members of the poppy family, the sepals fall as soon as the bud opens; but unlike other members, the sepals are three in number instead of two. Plants are avoided by domestic livestock, and their abundance in some places has been used as an indication of excessive rangeland use.

Flowering time: April to July   Poppy family

White-leaved Milkweed

Asclepias erosa Torr.

Also known as desert milkweed, this is the tallest of all milkweed species in the canyon country. The stems are mostly between 3 and 6 feet in height. It is a plant of warm to hot habitats, often among boulders at low elevations, from southern Utah and western Arizona south to southern California and northwestern Mexico. Kinds of native milkweed have been investigated as possible sources of domestic rubber. White-leaved milkweed is reported to be one of the most promising sources of rubber among plants native to the United States. The flowers, which are borne in simple umbels, rival those of the orchid family both in complexity of structure and in the manner in which they are pollinated. In both orchids and milkweeds the pollen is compacted into masses known as pollinia. The pollinia in milkweeds are attached in pairs by a miniature clip. The black clips are situated at the margin of the stigma. By accident, an insect must get its foot caught in the clip and pull the pollinia free. Once in another flower the pollinia must slip free from the foot of the insect directly on the stigma. Not many arrive successfully, but the number is indicated by one or two milkweed pods which mature on each plant.

Flowering time: June to September   Milkweed family

5

Sego-lily

Calochortus nuttallii T. & G.

Sego-lily, the state flower of Utah, is one of the most attractive members of the lily family. Flowers are 1 to 1½ inches wide and are borne on slender stems to about a foot tall. In some phases of the species, the flowers are pink to lavender in color. Members of the genus *Calochortus* (from the Greek words for *beautiful* and *grass*) are somewhat unusual for the lily family. Sepals differ vastly in size from the petals, and the texture of the sepals is more leaf- than petallike. Most other genera in the family have species with sepals and petals nearly alike. *Calochortus nuttallii* occurs from Oregon to North Dakota and south to California, Arizona, and New Mexico. The bulbs, which have the flavor of potatoes, were eaten by Indians and by pioneers. These are small, often less than ½ an inch in diameter, and are buried from 3 to 12 inches below the ground surface, commonly between rocks. Their use in times of stress can be justified only because they are an alternative to starvation. Their use in quantity for survival classes is not defensible. Attempts to cultivate this plant have not been successful. The future of the sego-lily lies in the perpetuation of wild lands.

Flowering time: April to July   Lily family

Douglas False-yarrow

Chaenactis douglasii (Hook.) H. & A.

The composite family is probably the largest single plant family, containing more than 20,000 species. It is remarkable that such a large family should have so few economically important species. Lettuce, artichoke, safflower, and sunflower nearly exhaust the list of edible species. The family does contain plants which are eaten by both domestic animals and wildlife. The Douglas false-yarrow is one of these. Plants are grazed by deer, sheep, and cattle. Generally, the false-yarrows are not a conspicuous part of the flora. Despite this, these plants and their flowers possess an intricate beauty. The heads of tiny white flowers, surrounded by green bracts, appear like small cushions. Protruding stamens and forked styles give them a lacy appearance. The divided bluish green leaves contribute to the overall laciness. The stems are 10 to 12 inches high and develop from a biennial root. Individual heads are about ½ an inch wide. Distribution is from British Columbia and Alberta southward to California, Arizona, and New Mexico. The plants occur in many soil types from low to high elevations.

Flowering time: April to July   Composite family

Utah Thistle

Cirsium utahense Petrak

The thistles of the canyon country fall into two groups, native and introduced. The Old World introductions are most common in cultivated lands, but occasionally they are found elsewhere. In the arid Southwest the native thistles are common. They do not spread to cultivated regions, but they do tend to increase on rangelands under heavy grazing. Thistles are seldom eaten by domestic stock because of their spiny leaves and flower heads. Despite the formidable appearance of the plants, the flower heads frame an intricate beauty. The flowers are arranged in powderpuff fashion, and all have tubular corollas. After the heads are mature and the seeds have been dispersed by wind acting on the umbrellalike pappus hairs, the numerous bristles which surround the flowers in the head become evident. Thus, even the mature heads which lack seeds lend a softness to an otherwise prickly plant. The plants are from 2 to 4 feet tall and seldom branch below the inflorescence. Each head is 1 to 2 inches wide in full flower. The Utah thistle grows in Utah, north-western Arizona, and in Nevada. Mostly it grows in the desert shrub, pinyon pine, and sagebrush vegetative types.

Flowering time: June to August   Composite family

Cliff-rose

Cowania mexicana DC.

Cliff-rose is an important winter browse plant for wildlife and for domestic livestock. These shrubs are commonly 4 to 6 feet tall, but occasionally some specimens reach 15 feet in height. The bark is easily shredded, as that of some species of juniper. It was used by Indians in the Southwest for making mats and other items. The flowers are commonly ½ to ¾ of an inch wide and are usually white, but cream and yellow phases are known. Flowering begins in early spring, when the numerous buds open over a short period of time, resulting in a beautiful display. Thereafter, the flowering continues at a reduced rate for the remainder of the growing season, but the array of numerous flowers must again await the arrival of spring. The fruit consists of five long-tailed, plumose-hairy achenes in each flower. In evening or morning light the rays of the sun cause the tailed achenes to glisten and individual plants to stand out halolike among the surrounding vegetation. The shrubs grow on hillside and rimrock outcrops in the juniper-pinyon and mountain brush zones. The area of distribution is from Nevada, Utah, and Colorado south to California and Mexico.

Flowering time: April to September   Rose family

Low Cryptantha

Cryptantha humilis (Gray) Payson

The genus *Cryptantha* (from the Greek words meaning *hidden flower*) is of moderate size, with more than 100 species indigenous to western North America and western South America. The low cryptantha is typical of a large number of perennial white-flowered cryptanthas which occur in the western United States. The dense clusters of flowers and sharply hairy stems and leaves easily distinguish this genus from all others in the region. Characteristically, the flower center is yellow in color, and the tube of the corolla is about as long as the calyx lobes. The fruit consists of four nutlets which are ornamented with bumps and ridges. The nature of the fruit ornamentation, or lack of it, has been used as a primary characteristic in the determination of species. Low cryptantha is from 2 to 15 inches tall. The flowers are from ¼ to ⅓ inch wide. The plants occur in valleys and foothills in the cool desert shrublands and upwards, from central western Colorado westward to eastern California, northwestern Arizona, southern Idaho, and southeastern Oregon.

Flowering time: April to July   Borage family

Jimsonweed; Sacred Datura

Datura meteloides DC.

The species of *Datura* are easily recognized by their large, sweet-scented, trumpet-shaped flowers and herbage which smells like a wet dog. When in favorable sites, the plants grow in rounded clumps to 4 feet in width. The foliage is gray green in color, adding contrast to the pure white flowers. The flowers of this jimsonweed are from 5 to 7 inches long and about as wide. They open in the evening and close the following morning. They are pollinated by night-flying insects, principally by moths. The fruit is covered with prickles and is 1 to 1½ inches in diameter. Daturas are poisonous, containing various alkaloids, notably atropine, hyoscyamine, and scopolamine. Various plant parts have been eaten by peoples of the Southwest, even by children, to induce visions. The practice is dangerous because of dosage problems; death can result from even a slight overdose. Contact with the foliage results in dermatitis in some people. Thus, this beautiful plant is not recommended for ornamental planting, a use to which it is occasionally put. The name *jimson* is an English contraction of Jamestown, where early settlers had difficulties after eating leaves of the related *Datura stramonium* L., the Jamestown weed. *Datura meteloides* grows at the bases of cliffs along drainages in Colorado and Utah and southward to California, Texas, and Mexico.

Flowering time: May to October   Nightshade family

Desert Star-lily

Eremocrinum albomarginatum (Jones) Jones

Desert star-lily was first named by resident western botanist Marcus E. Jones in 1891. The basis of the name and description was a series of specimens collected by Jones at Green River, Utah, on 9 May 1890. Jones, a pioneer botanist, first named the plant in an established genus, *Hesperanthes*. Later, he recognized that the desert star-lily was uniquely unlike any other member of the lily family, and he named and described the genus *Eremocrinum* (Greek words for *desert* and *lily*) to contain this beautiful plant. Only the one species is included in the genus, and it is said to be monotypic. Distributional area includes those portions of the Navajo Basin in Utah and northern Arizona. Thus, the desert star-lily is endemic to the Navajo Basin. Plants are seldom over 10 inches high, and the flowers are about a ½ inch broad. Sandy soils and dunes in the desert shrub community constitute the most important habitat. The leaves appear above ground in late winter or early spring, and the flowers are borne in mid- to late spring. Seed is produced by early summer, and by midsummer the plants have retreated into the ground.

Flowering time: April to June   Lily family

Round-leaved Eriogonum

Eriogonum ovalifolium Nutt.

Except for a few species, growing mainly on shale barrens in the eastern United States, the genus *Eriogonum* is a western American genus. Along with a few closely related genera, this genus is unique in the buckwheat family in lacking sheathing stipules at its nodes. *Eriogonum* is maintained in the buckwheat family because of flower structure, fruit, and seed characteristics. The genus is a large and complex one, with more than 200 species. Round-leaved eriogonum is characterized by having a leafless stem with a subglobose cluster of flowers situated at its apex. The flowers arise in a cup-shaped structure which simulates a calyx. The floral parts consist of six petallike segments which are ordinarily designated as sepals. The flower color varies from white to pink or yellow. The height of the plants is from 4 to 10 inches or rarely more. They grow in compact clumps and frequently produce several flowering stems to a clump. Thus, the plants are showy. They grow in sagebrush and mountain brush vegetative types and downward to the juniper-pinyon woods and desert shrublands. The species grows from Washington and Alberta southward to California, Arizona, and New Mexico.

Flowering time: April to June   Buckwheat family

10

Desert-thorn; Tomatilla

Lycium pallidum Miers

Several native species of desert-thorn occur in the hot desert areas of the southwestern United States. Only a few extend into the cool desert portions of the Four Corners region. Chief among the latter is *Lycium pallidum*. This spiny shrub grows to a height of 2 to 5 feet and flowers profusely during early spring and often after summer rains also. The flowers are from ½ to 1 inch in length. The plants grow along drainages near the margin of the range and over the general terrain in the hot deserts. Often, near the northern limits of the species, this desert-thorn is found associated with prehistoric Indian villages, possibly indicating its survival for long periods of time after being carried to the villages as food for the occupants. The fruit of desert-thorn was eaten both fresh and dried, often mixed with clay by Indians of the Four Corners area. The berries resemble small tomatoes, hence the name tomatilla. Flowers of this plant resemble those of the wild tobacco species in the region. Both tobacco and desert-thorn are in the same family. *Lycium pallidum* occurs from southern Colorado and Utah southward to southern California, Mexico, and western Texas.

Flowering time: April to June   Nightshade family

White Evening-primrose

Oenothera caespitosa Nutt.

The fragrant flowers of white evening-primrose open in the evening and remain open until the morning of the following day. During this time interval the petals are a pure white. The next evening the flowers open again, but this time they are pink in color; and when they close on the morning of the third day, not to open again, the petals fade to a dark pink color. Thus, it is not unusual for a given plant to bear white and pink flowers at the same time. The plants lack elongate stems; they have flowers and leaves borne closely together. Older plants tend to be hemispheric in shape and can bear few to many flowers during a single season. The blossoms are 2 to 3½ inches wide. They are conspicuous in late evening and early morning. In the daylight hours the flowers are tightly closed and quite inconspicuous. The plants grow in the desert shrublands and upwards to the mountain brush zone. The species is known in several forms from southwestern Canada southward to California, Arizona, and New Mexico. Several varieties of the white evening-primrose are known. They differ in flower size, leaf shape, and amount and type of hairiness.

Flowering time: April to August   Evening-primrose family

Hairy Evening-primrose

Oenothera trichocalyx Nutt. ex T. & G.

The stems of the hairy evening-primrose grow separately or in clumps to about 15 inches tall and wide. This evening-primrose is similar to the pale evening-primrose, *Oenothera pallida* Lindl, with which it grows through much of its range. The pale evening-primrose lacks long hairs on the flower buds; otherwise the flowers are very similar, being about 1 to 2½ inches wide. In both, the flowers open in evening and close the following morning. The evening-primroses are pollinated by night-flying insects, primarily by moths. They are parasitized by the green horn-worm, which is the larval phase of the sphinx moth. This is another example of a symbiotic relationship between animals and plants. The moth serves as the pollinating agent, and the plant serves as food for the moth. Sandy drainages, slopes, and sand dunes are their most common habitats. Hairy evening-primrose occurs from Wyoming southward through Colorado and Utah, while pale evening-primrose is distributed from Washington and Idaho southward to Arizona and New Mexico.

Flowering time: April to August   Evening-primrose family

Common Chokecherry

Prunus virginiana L.

Common chokecherry is a shrub or small tree which grows to about 20 feet in height. It occurs in shaded sites, moist hillsides, and in mountain brush, aspen, and lower spruce-fir vegetative types. The flower clusters appear along with the leaves and form an exquisite display, even though the individual flowers are small. The clusters of flowers are up to about 3 inches long and almost an inch wide. The foliage and other plant parts contain a substance (a cyanogenetic glucoside) which yields prussic acid (hydrogen-cyanide) in chemical breakdown; thus, under certain circumstances, the chokecherry is a poisonous plant. The cherries are less than ½ an inch in diameter, and the pit comprises most of the bulk. The thin fruit coat is astringent and somewhat bitter. Despite this fact, the fruits are eaten by birds and animals, and they are used in making excellent jellies and syrups. Indians used them, both fresh and dried, as a source of food. The common chokecherry is widely distributed in North America. It occurs in the eastern states, westward through most of the western states, and northward as far as Liard Hot Springs in northern British Columbia.

Flowering time: April to June   Rose family

Small-flowered Sandpuff

Tripterocalyx micranthus (Torr.) Hook.

Members of the genus *Tripterocalyx* are closely related to those of the genus *Abronia* (see p. 2). By some botanists, the two genera are combined; the name then becomes *Abronia*. Members of the two genera differ in that *Tripterocalyx* species are annuals with succulent (fleshy) stems that bear 2 to 4 very broad and thin, transparent, or translucent, conspicuous wings around the fruit body. The flowers of small-flowered sandpuff are slightly more than ½ inch long and less than ¼ inch wide. The clusters are dainty and attractive despite the small size of the flowers, which vary in color from greenish white to pink. The winged fruits take on a reddish hue and are more showy than the flowers. They are either sticky-hairy, as in many species of *Abronia*, or they are entirely lacking in hairs. A second species, *Tripterocalyx wootonii* Standl., is present in the canyon country. It differs in having bright pink flowers about an inch long and ½ an inch wide. Small-flowered sandpuff occurs from North Dakota west to Montana and south to Nevada, Arizona, New Mexico, and Kansas. Wooton sandpuff is known from Utah, Colorado, Arizona, and New Mexico.

Flowering time: April to June   Four-o'clock family

Narrow-leaved Yucca

Yucca angustissima Engelm.

The yuccas were possibly the singly most important non-cultivated plants to Indians of the Southwest. The buds, young flowers, and tender growing stalks were eaten by Indians, both raw and cooked, and the leaves were chewed. Sometimes the leaves were torn into longitudinal strips used for tying bundles and for other uses, especially in construction. Leaf strips were also plaited into sandals and mats. Cleaned fibers from yucca leaves were twisted into cord and small ropes, and these were worked into sandals and other items of fine quality. In most species of yucca the fruit matures into a dry capsule, and the narrow-leaved yucca is no exception. However, in the datil yucca, *Yucca baccata* Torr., the fruit is a thick and fleshy, pendant berry which is sweet when ripe. (The fruit of the datil yucca provided one of the real treats for Indians when the nearest ice cream parlor was a thousand years away.) Early pioneers used the roots of yucca as a source of soap, since they contain saponins which have soaplike qualities. The saponins function as laxatives, also. Narrow-leaved yucca is ordinarily less than 5 feet tall and grows well in sandy soils in southwestern Colorado, southeastern Utah, and northwestern New Mexico to northeastern and north-central Arizona.

Flowering time: May to June   Lily family

Kanab Yucca

Yucca kanabensis McKelvey

The yucca plants have been known by several common names, among them soapweed and Spanish-bayonet. The Kanab yucca is characteristic of a series of species with short leafy stems and slender leaves which grow in the Southwest. It has, however, a very restricted distribution, occurring only in south-central and southwestern Utah and northwestern Arizona. Plants with restricted range are known as endemics, and the Kanab yucca is a rather narrow endemic. It is closely allied to the Harriman yucca (*Yucca harrimaniae* Trel.), narrow-leaved yucca (*Yucca angustissima* Engelm.), Utah yucca (*Yucca utahensis* McKelvey), and the Toft yucca (*Yucca toftiae* Welsh), most of which have a broader distribution than does Kanab yucca. This species is abundant on the Coral Pink Sand Dunes near Kanab, Utah. It is easily recognized by its very long leaves and by the flower stalks which reach to 9 feet in height. The flowers are typical of those of most yuccas. They are pollinated by small moths that gather pollen into a mass and push it into the tube of the stigma. The moth larvae derive benefit by eating one or two rows of the seeds which are in six rows in each capsule. This is a classic example of a mutualistic symbiosis between a flowering plant and an insect.

Flowering time: May and June   Lily family

Panicled Death-camas

Zigadenus paniculatus (Nutt.) Wats.

The death-camas (also known as sand-corn) grows from an onionlike bulb. The plants are poisonous to livestock, especially to sheep. Cattle and horses seldom eat death-camas, and this probably accounts for their not being poisoned as often as sheep. The poisonous substance is found in all parts of the plant. During periods of near starvation early settlers in the Southwest occasionally mistook the bulbs of death-camas for those of onion, camas, brodiaea, or of sego-lily or mariposa lily with disastrous results. The plants of panicled death-camas lack the onion odor, have numerous flattened leaves arranged in three ranks, and produce a simple or branching cluster of tiny cream- to white-colored flowers. These features set the death-camas apart from the other edible plants with which they grow. The plants are commonly from 1 to 2 feet tall when in full flower. The individual blossoms are about ¼ inch wide. The panicled death-camas grows in the sagebrush and mountain brush zones from Washington to Montana and south to California, Arizona, and New Mexico.

Flowering time: April to June   Lily family

17

Trailing four-o'clock
(*Allionia incarnata* L.)
forms branches from the
summit of a perennial
root. The stems radiate
from the root to form a
circular mat on the surface
of the ground up to 4 feet
in diameter or more. The
flowers are ½ to ¾ of an
inch wide and occur in
shades of pink and rose
purple or, less commonly,
white. The species is, for
the most part, an
inhabitant of warm
deserts, where it grows in
sandy and gravelly soils
with desert shrubs and
grasses. Trailing four-
o'clock occurs in west-
central Colorado, southern
Utah, and southward to
Mexico and South
America.

Hooker Onion

Allium acuminatum Hook.

The onion, chives, leeks, and garlic of commerce belong to the genus *Allium*. Onions are probably among the most ancient of the cultivated plants in the Old World, and they are known to have been eaten by prehistoric peoples in the western hemisphere as well. Plants of the genus *Allium* have an odor that easily distinguishes them from most other plants, an odor of onion or of garlic. All species are considered to be edible, but the flavor and odor are offensive to some people even though the plants are nutritious. When large quantities are eaten by dairy animals, the milk develops a strong and distasteful flavor. Wild onions constitute an important item of diet in survival training programs, but if one were truly lost and had been forced to survive by eating wild onions, his rescuers might be reluctant to rescue him. Certainly onions improve the flavor of bland foods and can be used to season foods in outdoor cooking. The Hooker onion grows from a deeply placed bulb. The flowering stem is commonly 5 to 10 inches tall, and the flower cluster is 1 to 2 inches wide. The plants occur in shrublands and open woods at middle elevations in much of the western United States.

Flowering time: April to July   Lily family

Nevada Onion

Allium nevadense Wats.

Allium is a genus of moderate to large size with about 300 species widely distributed in both the Eastern and Western hemispheres. The name is derived from a Greek word meaning *to avoid* (on account of the odor or flavor or both). There are more than a dozen species of onion in Arizona alone and more than 40 in California. The species are distinguished on the basis of technical features involving the nature of bulbscales, number and type of leaves, and shape and form of the ovaries. There are very few species with solitary leaves; most have at least two leaves per flower stalk. The leaves are borne at the apex of the deeply buried bulb, and the scape with its umbel of flowers extends from the sheathing leaf base. Patterns of the outer bulb scales are intricate. These provide a kind of species fingerprint by which unknown specimens can be identified. In Nevada onion the naked flower stalk, or scape, is generally less than 8 inches tall. The umbel consists of mostly 12 to 30 flowers which vary in color from pale pink to white. The species occurs on a variety of soils from southeastern Oregon and southern Idaho south to Arizona and California and east to west-central Colorado.

Flowering time: April to June   Lily family

Crescent Milkvetch

Astragalus amphioxys Gray

There are more than 350 species of *Astragalus* in North America, with most of them confined to the western states. Utah alone hosts more than 100 species, and the representatives of this group form complex and very difficult assemblages in all western states. The genus is represented by a great many species in Eurasia, and there are some in South America also. Certainly *Astragalus* is one of the largest genera of flowering plants. Variation in the genus is very great, and the species included in this book represent only a small portion of the total (pages 22, 47, and 72). *Astragalus amphioxys* belongs to a group of species with short stems, with most of them occupying open bare ground at low to middle elevations. They form a distinctive group, differing from each other in degree of hairiness and kind of hairs, number of leaflets per leaf, flower size and color, and especially in features of the pod. The flowers of the crescent milkvetch are from about ½ to 1 or more inches long. The pods are crescent-shaped at maturity and are deciduous from the plant. Finally, the pods split open at the apex and the seeds fall free. The fresh pods are relatively heavy, weighting the inflorescence in such a manner that ultimately the pods recline on the ground.

Flowering time: March to June (October)   Legume family

Milkweed Milkvetch

Astragalus asclepiadoides Jones

The milkweed milkvetch is one of the preeminently singular species in the genus *Astragalus*. The leaves of all other species are pinnately compound (i.e., divided into separate leaflets); or if simple they are very narrow. The leaves of the milkweed milkvetch are broad; even though they are alternate, they closely resemble those of the milkweed, *Asclepias cryptoceras* Wats., with which they grow. The cryptocera milkweed has reclining stems and other features which separate it easily from this most unusual milkvetch, but the leaves are similar in size, shape, and texture. The milkweed milkvetch is a primary indicator of selenium in the soils in which it grows; in this respect it is similar to yellow milkvetch (page 47). The plants are from 6 inches to 2 feet tall and grow in clay soils in western Colorado and in central to eastern Utah. In some seasons they are common along roadsides where they receive extra water from pavement runoff. The stems grow singly or in clumps of several. They are commonly erect to ascending, and the inflated pods tend to stand at a steep angle from the flower cluster in which they are borne.

Flowering time: April to June   Legume family

21

Thompson's Woolly Locoweed

Astragalus mollissimus Torr.

The plant figured here is Thompson's variety of the woolly locoweed, or *Astragalus mollissimus* var. *thompsonae* (Wats.) Barneby. The species is rather widely distributed in the plains states, in the Southwest, and in northern Mexico. It is a highly variable entity, and some eleven distinctive varieties are known. The Thompson's woolly locoweed occurs in eastern and southern Utah, southwestern Colorado, northwestern New Mexico, northern Arizona, and southern Nevada. The plants are from 4 to 10 inches tall or sometimes more, and the flowers are about an inch long. The varieties of woolly locoweed are all toxic, especially to horses, and in a lesser degree to sheep, cattle, and goats. The poisonous agent is thought to be an alkaloid called locoine. In some other species of *Astragalus*, the poisonous substance is the element selenium, which is concentrated in the plant tissues. The genus is a huge one and has numerous nonpoisonous members. Some important forage species occur in this genus, especially in the Old World. This locoweed is one of the earliest of spring flowering plants in the canyon country.

Flowering time: February to June   Legume family

Bent Mariposa

Calochortus flexuosus Wats.

In arid lands, possibly the showiest flowers of the lily family belong to the species of *Calochortus*. The large tuliplike blossoms borne on slender, bluish green stems and reduced leaves, which are often withered by flowering time, are definitive features of the group. In bent mariposa the stems, which are 6 to 18 inches long, commonly do not support their weight, but instead they recline on the ground or on other vegetation and assume a zigzag or twisted posture. The flowers are 1 to 1½ inches long and about as wide. The petals and stamens wither and fall away, leaving the ripening ovary alone at the apex of the flower stalk. In time the ripened ovary (a capsule) splits apically along predetermined lines, and the black waferlike seeds are exposed to the drying air. The capsule opens wide with successive drying, and the seeds are freed by air currents. The plants grow in sandy or silty soils in the desert shrub zone of southwestern Colorado and southern Utah and southward to eastern California and central Arizona. The bulbs of this species are edible also, but they are as difficult to harvest as are those of the sego-lily (page 7).

Flowering time: April to June   Lily family

Indian Paintbrush

Castilleja chromosa A. Nels.

There are many species of paintbrush in North America. All are brightly colored, commonly with shades of red or purple, but some are yellow to cream. The showy part of the paintbrush is not the flowers but the bracts which subtend them. The calyx is brightly colored also, but the corolla, which is mostly hidden by the enclosing calyx, is commonly greenish or whitish except for the apex. The paintbrushes are root parasites on other plants, but evidently they manufacture much of their own food. The paintbrush figured here grows at low elevations through most of the Four Corners states. It comes in flower during early spring; because it is so conspicuous, it is a favorite of collectors and photographers. The plants are from 4 to 12 inches tall. They grow in sandy or gravelly soils in much of western North America. The plants occur in desert shrublands, pinyon-juniper forests, and in the sagebrush and mountain brush zones. The presence of glossy hairs in the inflorescences of many species adds to the attractive appearance of the plants.

Flowering time: March to June   Figwort family

Bull Thistle

Cirsium vulgare (Savi) Tenore

The bull thistle is a pestiferous biennial weed of Old World origin which is widely distributed in North America. The plants are most common in agricultural regions, where they grow in disturbed soils along roads and ditches and in pastures. However, since each of the numerous seeds produced by a single flower head is equipped with an umbrellalike pappus, they are effectively dispersed by wind; so the plants might be expected to occur anywhere. Bull thistle grows to be from 2 to 6 feet tall. The flower heads are 1 to 2 inches wide. In spite of the spiny leaves, heads, and stems, which protect the plant from most grazing animals, the plants are edible. With spines removed, young stems can be eaten either as potherbs or uncooked. The name thistle has been applied to plants belonging to several different genera. The name has been given to the genus *Onopordum*, which contains the Scotch thistle, inseparably associated both in song and story with Scotland. *Onopordum* is likewise a weed of disturbed roadsides and cultivated land in the western states. It has winged stems like bull thistle but is almost white in color due to a covering of cobwebby hairs.

Flowering time: June to September   Composite family

Shooting-star

Dodecatheon pulchellum (Raf.) Merrill

The nodding flowers of the shooting-star species, with bright pink reflexed petals, are some of our most striking and beautiful plants. Members of the genus *Dodecatheon* commonly occur in moist meadows at middle and high elevations, but they grow at low elevations in some wet seeps and hanging gardens in the canyon country. The naked flowering stems, or scapes, are from 6 to 15 inches tall, and each bears a cluster of few to several flowers at its summit. Flowers are from ½ to 1 inch long and are themselves reflexed at the ends of the pedicels. There is much variation in plant size, hairiness, and other features of this widely distributed species. Indeed, the plants from the seeps and hanging gardens of the Southwest have received several names because of the morphological differences they exhibit. Arrangement of stamens in *Dodecatheon* is in contrast to that found in most other families of flowering plants. The stamens of shooting-star are situated in front of the petals, as is characteristic of the primrose family, instead of between the petals, as is characteristic of most other families. Thus, the character of the stamens is useful in distinguishing members of this family. Shooting-star is distributed from Alaska eastward to Pennsylvania and south to Mexico.

Flowering time: April to August   Primrose family

Engelmann Hedgehog Cactus

Echinocereus engelmannii (Parry) Rumpler

The plants of Engelmann hedgehog cactus occur as solitary slender barrels or in clusters of few to many. Flower buds arise within the tissues of the stem adjacent to the point of attachment of spines some distance below the tip of the stem. The buds become visible only when they break through the epidermis. The flowers are waxy. They open in mid-morning and close at night, lasting for several days. The fruit has large, readily detached clusters of spines and is thin-fleshed but juicy and edible. The fleshy tissue of the thick stems is mucilaginous and bitter. It does not offer much hope of water for thirsty travelers but might be eaten as an emergency food. The plants are seldom more than 8 to 10 inches tall and the flowers are 1 to 2 (rarely 3) inches in diameter when fully open. Through most of its area of distribution, Engelmann hedgehog cactus is not abundant. Individual stems or clusters are likely to be widely spaced in gravelly or sandy soils where they occur. The area of distribution is from southern Utah through Arizona and southern Nevada, also south to Mexico. Like many cacti, this plant is being reduced by collectors who sell them commercially.

Flowering time: April to June   Cactus family

Helleborine

Epipactis gigantea Dougl. ex Hook.

The helleborine orchid, like most other orchids, requires a
moist place in which to grow. Since most of the southwestern
United States is arid, the number of orchid species in the
region is very limited. Most of those which do occur in the
vicinity grow in wet sites at middle and higher elevations. The
number of appropriate habitats at lower elevations is very
small. Mostly the species occur in the mass of vegetation
around springs and seeps and in the hanging gardens. Plants
of helleborine grow from 5 to 15 inches tall. The flowers are
from ½ to ¾ inch wide. Individually, the bilaterally
symmetrical flowers are very showy, but the plants tend to
blend into the dense vegetation in which they grow and are
easily overlooked. In many of the sites where helleborine
grows, a second type of orchid occurs also. It is the few-
flowered bog orchid, *Habenaria sparsiflora* Wats., which has
small, delicate, sweetly scented white flowers. The helleborine
occurs from British Columbia and Montana southward to
California, Arizona, and Texas.

Flowering time: April to July   Orchid family

Butterfly-weed

Gaura coccinea Nutt. ex Pursh

The butterfly-weed is easily overlooked because the flowers are small and are not colored brightly. It is like many other members of the evening-primrose family in that it produces flowers which are one color upon opening and another color, usually pink, upon fading. The plants grow to about 15 inches in height, and the flowers are from about ¼ to ½ inch wide. In most members of the family the fruit opens at maturity and the numerous seeds fall free. The butterfly-weed is exceptional because it has fruits with one to four seeds and does not open at maturity. The fruit is shed as a unit, the seeds germinate within the fruit coat, and finally the seedling penetrates through the seed coat and fruit covering. Floral parts are in fours or multiples of four, as is characteristic for most genera in this family. Commonly there are four each of sepals and petals, eight stamens, and a four-carpeled, inferior ovary. The plants grow in grasslands and shrublands at lower elevations from southern Canada to Mexico. In the canyon country they occur in southwestern Colorado, southern Utah, Arizona, and New Mexico.

Flowering time: April to September
Evening-primrose family

Scarlet Gilia

Gilia aggregata (Pursh) Spreng.

Scarlet gilia (also called skyrocket) is one of the showiest wild flowers in the canyon country. The plants are biennials, with the flowering stem elongating from a basal rosette of leaves in the second growing season. The plants are from 6 inches to 2 feet tall. The slender flowers are from ¾ to 1½ inches long and are commonly red. However, there are forms with pink, orange, and white flowers. The plants have a heavy, slightly skunklike odor which also occurs in some other members of the phlox family. Scarlet gilia grows throughout the canyon lands from low to high elevations in most vegetative types. It is a very common species in western North America. Scarlet gilia is eaten by grazing animals and is pollinated by hummingbirds and by moths, both of which are equipped to reach the nectaries at the bases of the tubular flowers. The greatest display of flowers occurs during early springtime, but the plants will continue to grow and put forth new buds for most of the growing season, often until after frost in late autumn. The startling beauty of a fresh and bright scarlet flower clinging precariously to a dried and scarred plant in the clear cold air of autumn is sufficient reward for having searched and for being willing to stop and admire.

Flowering time: April to November   Phlox family

Spiny Hop-sage

Grayia spinosa (Hook.) Moq.

This book is dedicated to a presentation of flowers, their variations, and especially their beauty. Some plants, and the spiny hop-sage is one of them, have tiny and inconspicuous flowers which lack showy corollas. Despite this apparent shortfall, the plants put on a remarkably beautiful display of form and color. Attached below the ovary in female flowers are two bracts. These enlarge and enclose the developing fruit, and it is the enlarged bracts which take on hues of red, orange, or purple. Some mesa tops in portions of the canyon country where spiny hop-sage is abundant take on the aspect of the color of these plants, changing the entire scene to a shade of red-purple. Despite the spiny branchlets which serve as protection, this plant is palatable to both livestock and to wild animals. Cattle, sheep, and horses eat the slender stems, mainly in winter, and because of the partiality of livestock, some herders have called this plant "apple bush." Spiny hop-sage occurs on sandy, well-drained soils from Washington east to Wyoming and south to California and Arizona. The plants are members of the semidesert shrub community. They are low shrubs, mostly from 2 to 3½ feet tall.

Flowering time: April to June   Goosefoot family

Northern Sweetvetch

Hedysarum boreale Nutt.

In general aspect of foliage and flowers, the sweetvetches resemble the milkvetches and locoweeds. The short wing petals and abruptly angled keel petals are diagnostic features which distinguish the sweetvetches in flower. The nature of the sweetvetch fruit also makes the plants easy to identify. The fruit is a loment, a kind of legume or pod which breaks into one-seeded segments at maturity. In other legume genera in western North America the fruit is ordinarily a pod which splits along predetermined lines at maturity, letting the seeds fall free. The plants grow from 1 to 2 feet tall, and the flowers are from ½ to 1 inch long. The species occurs from Alaska and the Yukon southward to Nevada, Arizona, and Oklahoma. The roots of some species of sweetvetch are eaten by Eskimos and Indians, but those of the northern phase of *Hedysarum boreale*, the ssp. *mackenziei* (Richards.) Welsh, were reported to produce illness when eaten by members of Sir John Franklin's expedition to the Polar Sea in 1817. However, they ate the roots of the alpine sweetvetch, *Hedysarum alpinum* L., without discomfort. The northern sweetvetch grows in the desert shrub, pinyon-juniper, ponderosa pine, and mountain brush zones of the Southwest.

Flowering time: April to July   Legume family

30

Morning-glory Heliotrope

Heliotropium convolvulaceum (Nutt.) Gray

The flowers of morning-glory heliotrope are very attractive. They are about ½ an inch long and as wide or wider. They open in the late afternoon and are sweetly scented. The limb of the corolla is white to pinkish except for the yellow marking around the center. The plants branch profusely when water is abundant, and it is not uncommon to see rounded clumps 4 to 10 inches wide with numerous flowers simultaneously open. The usual place of growth is in sandy soils, often in sand dunes. In the canyon country of the Southwest, the sand grains are coated with a reddish covering, giving the soils a reddish or pinkish cast. This background accentuates the contrast between plant, flowers, and soil. The result is one of unmatched beauty. Even in death the stems of this heliotrope continue to present a display. The plant parts bleach to a straw white color and persist for a year or more. Following a rain in late autumn, the droplets hang like jewels on the skeletonized branches of this and other plants of dunes. The area occupied by morning-glory heliotrope includes Nebraska west to Utah and south to Arizona, Mexico, and Texas.

Flowering time: March to October   Borage family

Skeleton-plant

Lygodesmia grandiflora (Nutt.) T. & G.

Most members of the composite family in the western United States have yellow flowers. Thus, the bright pink flowers of skeleton-plant are somewhat anomalous. It is also unusual for such large heads to be borne on a relatively small plant. Skeleton-plant is in the same group of genera as the dandelion, *Taraxacum officinale* Weber, and has milky juice also. The plants are from 4 to 15 inches tall and commonly branch from near the base. The heads of flowers are from 1 to 1½ inches long and about as wide. Fresh plants have been used along with others having milky juice to stimulate milk flow in nursing mothers. The use of plants having parts resembling human anatomy or secretions to treat ailments of those parts or secretions was common in Europe during the Middle Ages; its practice was known as the doctrine of signatures. In other societies such uses have been designated as sympathetic magic. Surprisingly, some of the treatments were efficacious, and even more surprisingly, few people failed to survive the treatments. Skeleton-plant grows in sandy and gravelly soils in grassland, desert shrublands, and in sagebrush communities from Idaho and Wyoming south to Arizona and New Mexico.

Flowering time: May to June   Composite family

Eastwood Monkey-flower

Mimulus eastwoodiae Rydb.

There are two species of monkey-flower in the canyon country that have bright red flowers. The one is figured here, and the other is the crimson monkey-flower, *Mimulus cardinalis* Dougl. Evidently both occupy similar habitats in the gardens which develop on moist cliffs or in alcoves in the drainages of the Colorado River canyon. Eastwood monkey-flower is common in the hanging gardens in southeastern Utah and northeastern Arizona, occupying the uppermost stratum of vegetation or growing intermixed with other species on the moist slope at the lower edge of the garden. The crimson monkey-flower occurs on the wet rocks in southwestern Utah (e.g., on the weeping rocks in Zion Canyon) and in the region of Grand Canyon in Arizona. The crimson monkey-flower tends to have upright stems and larger flowers than those of the Eastwood species, which tends to have stems that are appressed to the sandstone. The red corolla of Eastwood monkey-flower is from 1½ to 2 inches long or more. It is one of the few truly red-colored flowers of summertime in the canyon country. The flowers resemble those of the beardtongue species, but flowers of *Mimulus* species possess only four stamens.

Flowering time: June to August   Figwort family

Wild Four-o'clock

Mirabilis multiflora (Torr.) Gray

The four-o'clocks are unusual in that they have showy flowers which lack petals. The colored whorl of floral parts which simulate petals are actually sepals. To make matters more confusing, the flowers are subtended by green bracts which simulate sepals. Despite their unusual nature, the flowers are truly beautiful. The stems are heavy and tend to recline on the ground. The number of stems varies from few to many, and their length is from a few inches to 3 feet. In full flower, the entire plant, in clumps to 5 or 6 feet wide, often appears rose pink to purplish red due to the large number of blossoms. The flowers are from ¾ to 1½ inches wide. Plants of this species are almost definitely associated with juniper-pinyon woodland. In forests of those species in portions of the Southwest the wild four-o'clock grows only under trees of juniper or pinyon. The leaves appear in early spring, even in very dry seasons, but flowering is profuse only when moisture during springtime is abundant. The plants occur from southern Utah and southwestern Colorado south to Mexico. Flowers open in early evening and remain open throughout the night. By mid- to late morning they close for the day. Pollination is by night-flying insects.

Flowering time: April to September   Four-o'clock family

Hedgehog Prickly-pear

Opuntia erinacea Engelm. and Bigel.

This prickly-pear is similar in size and other features to *Opuntia polyacantha* (page 61). The padlike joints of the stem are from 3 to 6 inches long and are roughly oval in outline. In addition to being armed with coarse, sharp spines, there are clusters of exceedingly tiny spines called glochids at the base of the larger spines. If one handles prickly-pears, it is fairly easy to avoid the coarse spines, but the glochids are something else . Once they penetrate the skin, they are difficult to remove without the aid of a magnifying glass and a high-quality pair of forceps. Additionally, clusters of glochids penetrate clothing and reappear in fingers or elsewhere when the clothing is handled, sometimes days or weeks following the encounter with the prickly-pear. The glochids are more of a nuisance than anything else, but they supply reason enough for most people to leave prickly-pears alone. The flowers of *Opuntia erinacea* are from 1½ to 3½ inches wide. The plants grow in a variety of soil types in the pinyon-juniper and desert shrub vegetative types. The range includes southwestern Colorado, southern Utah, southern Nevada, southeastern California, and Arizona.

Flowering time: May to July   Cactus family

Bearded Beardtongue

Penstemon barbatus (Cav.) Roth

The stems of *Penstemon barbatus* occur singly or as a few grouped together from a perennial root crown. They are from 15 inches to 4 feet tall, bearing flowers which are from 1 to 1½ inches long. The projecting upper lip and recurved lower lip with long yellowish hairs, combined with other features, easily distinguish this species from the Eaton beardtongue (page 37). These two species are almost singular in *Penstemon*, wherein few species possess red or scarlet flowers. The great majority of members of the genus display corollas that are some shade of pink, pink purple, blue, or purple. Several phases of *Penstemon barbatus* are known. Basic structure of flowers is the same in all, but they differ in minor ways. One has anthers with long yellow hairs, and another lacks hairs on the lower lip. Glands which secrete nectar are placed at the base of the tube on the inside of the flower. Hummingbirds and insects with especially long mouth parts are equipped to harvest the nectar and to serve as pollinating agents. Habitats occupied by bearded beardtongue include open to shaded sites from low to high elevations in pinyon-juniper, mountain brush, ponderosa pine, and spruce-fir vegetative types. The species is distributed from southern Utah and southern Colorado south to the central highlands of Mexico.

Flowering time: June to October   Figwort family

35

Eaton Beardtongue

Penstemon eatonii Gray

It is relatively easy to introduce Eaton beardtongue to cultivation. Seeds gathered in late summer can be planted in sandy soil in flower pots and placed outside to overwinter. Seedlings can be transplanted to plots of well-drained gravelly or sandy soil where they begin to flower in one to three years. Basal rosettes of leaves are formed that persist through the winter. Flowering stems elongate from the short stem produced in the rosette of the previous year. The plants are perennial, and unless shaded or watered too heavily they will persist for several years. A mature plant will produce several stems of flowers, blossoming over a period of two weeks or more. This beautiful flowering plant will provide many blossoms which will serve as natural feeders for hummingbirds and exotic types of insects. In mid- to late summer the capsules mature, and each capsule will contain a great number of tiny seeds. Capsules are often pigmented with red and are attractive even after the corollas have withered and fallen. Stems of Eaton beardtongue are commonly 2 to 3 feet tall, and the flowers from 1 to 1½ inches long. The plants occur from north-central Utah and southwestern Colorado to Arizona, southern Nevada, and southern California. They grow in desert shrublands, pinyon-juniper woods, mountain brush, and in mountain forests at high elevations.

Flowering time: February to June   Figwort family

Palmer Beardtongue

Penstemon palmeri Gray

Penstemon is a large genus of especially beautiful plants with more than 200 species. Mainly they are North American, with the major center of distribution located in the West. There is also one species known from Asia. The generic name is derived from the Greek words for *five* and *thread*, an allusion to the five stamens produced in the flowers. The stamens occur in two pairs, with the fifth or odd stamen lying in the throat of the corolla. In *Penstemon* species the fifth stamen is sterile (a staminode) and is commonly bearded — hence the common name of beardtongue. Among the most beautiful of the species in the canyon country is the Palmer beardtongue. Clumps to two feet tall or more and almost as broad stand out startlingly against a landscape of contrastingly colored sand and rock. The seemingly inflated corollas, more brightly colored guidelines, and the bright yellow beards of the staminodes set this flower apart as outstanding. Adding to the charm of this plant is a subtle fragrance which issues from its flowers. The expanded throat of the corolla and the pale color indicate that agents other than hummingbirds are responsible for most pollination. Even large bees can have access to the inside of the flowers. The plants grow in sandy soils from southern Utah to Arizona and California.

Flowering time: March to September   Figwort family

37

Long-leaved Phlox

Phlox longifolia Nutt.

The long-leaved phlox is a harbinger of spring. Its straggling stems and bright, relatively large flowers easily set it apart from most other species in the western states. The stems often exceed 4 inches in length; when they are protected beneath a woody shrub, they often grow to the height of the shrub and have a definitely woody stem below the shrub crown. Thus, a plant might be from 1 to 2 feet tall or even more. The tube of the flower is from ½ to ¾ of an inch long, and the limb is about as wide. Flower color is variable, but the plants are usually some shade of pink. Plants are palatable to livestock and to wildlife. This accounts for their diminutive size except where they are protected. There are some 50 to 60 species of the genus *Phlox*, almost all of them native to North America. The species are best developed in western North America. The name *Phlox* is a direct translation of the Greek word for *flame,* a reference to the brightly colored flowers of most species. Most of the species in the western states have low matted stems which are appressed to the surface of the ground. Thus, the long-leaved phlox is rather easily distinguished. It occurs in sandy and gravelly soils in desert shrub, pinyon-juniper, and sagebrush zones from British Columbia and Wyoming south to California, Arizona, and New Mexico.

Flowering time: March to June   Phlox family

Cave Primrose

Primula specuicola Rydb.

Not many travelers are fortunate enough to see the handsome cave primrose of the canyon country. It flowers very early, beginning as early as January in some years, and reaches a flowering climax in March or April. In recognition of the flowering time, the common name "Easter flower" has been applied in some regions. There are few primroses which occur in the canyon country. Those that do are commonly plants of high altitudes in mountains. Cave primrose is unique among them in occupying low elevations in this great arid land. The plants do not, however, have to tolerate the great desiccation of the region. Rather, they grow in the seeps and alcoves that mark canyon walls in massive sandstones. Here the cave primrose grows with a plant assemblage peculiar to the area. These green patches clinging to cliff faces seem foreign, for they often form the only real green vegetation in a landscape where the thin mantle of plant cover is gray, gray green, or even blackish in aspect. Plants are from 2 to 10 inches tall, and the flowers are ½ to ⅔ of an inch broad. They are restricted (endemic) to the wet seep walls in canyons of the Colorado in southeastern Utah and northern Arizona.

Flowering time: January to May   Primrose family

Canaigre

Rumex hymenosepalus Torr.

Canaigre is known by other common names; perhaps the most prominent is "wild-rhubarb." The leaves have been eaten as greens, either boiled or roasted, and the petioles cooked with sugar as is the rhubarb of commerce. The plants begin growth in early springtime while most other plants are still dormant. The stems arise from a deeply buried cluster of tuberous roots. The roots are rich in tannin, and because of this canaigre has been investigated as a source of commercial tannins. However, attempts at cultivation have not proved to be economical. Indians used the roots medicinally for treatment of minor ailments. Recently they have appeared on the market in the Southwest labeled as "ginseng," whose roots they vaguely resemble. The efficacy of the roots, if any, is unknown. As in spiny hop-sage, a member of a related plant family, the flowers of canaigre are not showy. The perianth consists of six calyxlike parts, usually green to red in color. The three inner perianth segments become enlarged and modified in fruit. In canaigre, these segments are up to half an inch long and broad, brightly red colored, and membranous in texture. It is from the membranous nature of the enlarged perianth segments that the scientific name is derived. The plants grow in sand dunes and other sandy sites from Wyoming and Utah south to northern Mexico.

Flowering time: March to April   Buckwheat family

Fishhook Cactus

Sclerocactus whipplei (Engelm. & Bigel.) Britt. & Rose

Sclerocactus is a small genus of southwestern cacti, occurring mainly in Utah and Arizona. The most distinctive feature of this genus involves the shape of the spines. At least one of the central spines in each cluster is recurved to form a definite hook from which the common name is derived. Attempts to handle this plant in any manner demonstrate the effectiveness of the hooked spines. It is possible to avoid the hooked and the adjacent straight spines until one tries to withdraw a finger and the needle-sharp end of the hooked spine penetrates the skin. One such encounter is sufficient to deter even the most enthusiastic naturalist. The fishhook cactus is the common small barrel cactus of the canyon country. The plants grow from 2 to 15 inches tall and up to 6 inches in diameter. The flowers are 1 to 2 inches long and open to about as wide. They occur in yellow to white, cream, and very pale pink phases as well as the typical pink to purplish forms. The petals possess the luminescent quality that is characteristic of cactus flowers generally, and they tend to dazzle and deceive both eye and camera lens. Plants are widespread in the desert shrublands and pinyon-juniper woodland of Utah, Colorado, and Arizona. They occupy soils which vary from clay and silt to sand and gravel.

Flowering time: April to June   Cactus family

Wire-lettuce

Stephanomeria exigua Nutt.

Wire-lettuce, as the name implies, is a relative of cultivated lettuce which belongs to the genus *Lactuca*. The stems are thin and wiry and the small leaves are bitter. Thus, they are not considered to be edible to man. They are eaten by grazing animals, but since wire-lettuce is seldom abundant it is not considered to be an important range plant. The flower heads of wire-lettuce closely resemble those of skeleton-plant (page 33) in color and general form, but they are much smaller (only ½ to ¾ of an inch long and about ½ inch wide). The plants are more delicate, with slender, multiflowered branchlets and more numerous flower heads. Like other annuals, the size of the plant is determined by climatic factors, but in any given year one can find plants from 4 to 15 inches tall, depending on the conditions of the local environment. This wire-lettuce grows in sandy to clay soils in desert shrub and pinyon-juniper zones throughout Wyoming, western Colorado, Utah, New Mexico, Arizona, and California. Members of *Stephanomeria* belong to that group of the composite family with ray flowers only and with plant parts having milky juice.

Flowering time: April to September   Composite family

Salt-cedar; Tamarix

Tamarix ramosissima Ledeb.

Tamarix is a handsome shrub or small tree, especially when it is in full flower. The combination of spikes of tiny pink flowers and slender, juniperlike branchlets produces an illusion of softness. Despite the striking nature of tamarix, it is a pestiferous weed which was introduced from the Old World as an ornamental plant. The seeds, which are produced in abundance, are each equipped with a tuft of hairs; they are easily dispersed by even slight breezes. Following introduction the plants spread rapidly. They reached the upper Colorado River drainage sometime after 1910 and in the next twenty years occupied essentially all available sites along the rivers, streams, and seeps of the region. Tamarix has successfully competed with existing streamside vegetation, and in many regions it has replaced the native plants. In addition to being less palatable to domestic livestock and wildlife than native vegetation, Tamarix wastes water; and in a region where water is the key to economic survival, a plant which wastes water is a detriment. The shrubs grow to a height of about 20 feet and form almost impenetrable thickets along the major rivers in the canyon country.

Flowering time: March to September   Tamarix family

The century-plants are
mostly distributed in areas
south of the canyon country,
and it is somewhat surprising to
note the very long flowering
stems of the two species
which grow in the Grand
Canyon and Virgin River
areas of Arizona and Utah.
Kaibab century-plant (*Agave
kaibabensis* McKelvey) is
known only from the Grand
Canyon region of north-
central Arizona. It is a plant
with very long sentinellike
flower stalks arising from a
solitary rosette of ascending-
spreading basal leaves.
Nevertheless, the stalks —
whether fresh and flowering,
in fruit, or dead and bleached
— mark the sites inhabited
by these plants with flaglike
precision. They grow
vegetatively for several years
before the flower stalk
begins to grow, when the
growing point of the plant is
used up in the inflorescence
and the rosette dies.

Yellow Columbine

Aquilegia chrysantha Gray

Yellow columbine occurs in the canyon and plateau country of southern Colorado, New Mexico, and Arizona and southward to northern Mexico. The large, long-spurred flowers with canary-yellow petals make this one of the most attractive plants in the genus. The flowers are from 1½ to 3 inches long. Nectar is produced by glands within the swollen tip of each spur. Pollinating agents are ordinarily butterflies or moths, whose coiled mouth parts unroll to sufficient length to allow them to reach the nectar. The fruit consists of five separate follicles which open along one side, each bearing numerous tiny seeds. The yellow columbine grows through a wide range of elevations from low semidesert sites to cool alpine regions. The genus *Aquilegia* (from the Latin name for *eagle*, a reference to the clawlike spurs) contains some 70 species. They are widely distributed in the Northern Hemisphere. There are other species known from the canyon country (see page 70), but most of them occur in woods at high elevations.

Flowering time: May to August   Buttercup family

Rough Milkweed

Asclepias asperula (Decne.) Woodson

The flowers of rough milkweed form yet another type in the great range of variation in flowers of members within the genus *Asclepias* (pages 5 and 47). The petals of the flowers in rough milkweed are green in color. Only the five lobes of the corona are brightly colored. Whitish tips of corona and stigma form a contrast to the flower parts, providing a targetlike appearance. Insects which are attracted by the unusual display take nectar from glands at the base of the corona, and at least the larger insects carry an occasional mass of pollen (pollinium) to the flowers on adjacent plants. Small insects sometimes get caught in the clip mechanism of the pollinia; lacking sufficient strength to pull the pollinia free from the anthers, the insects remain trapped until they die. Flowers are only about ½ an inch broad. They are borne in clusters 2 to 3 inches wide. Stems of rough milkweed occur singly or in clumps of several. Some species of milkweed are poisonous to livestock; rough milkweed is suspected of being poisonous. The plants grow in many soil types at low to middle elevations in desert shrub, juniper-pinyon, and mountain brush vegetative types from Kansas, Colorado, and Utah south to Arizona and Mexico.

Flowering time: April to August   Milkweed family

45

Pleurisy-root

Asclepias tuberosa L.

In the scheme of nature in the Southwest there are very few species of plants with orange-colored flowers, and this is especially true in the canyon country. Species of *Sphaeralcea* (page 64) are the main ones with orange flowers. Thus, it is all the more surprising to encounter a plant of pleurisy-root, also known as butterfly-weed, in some remote section of the arid southwestern states. The startling contrast of orange against green foliage, or against a backdrop of sand or stone, often makes one stand in awe. In the orchid family, where flower structure is exceedingly complex also, there is much variation in the form of flowers, especially in size and shape of the various parts. Structure of milkweed flowers is also very complex, but the flowers are remarkably similar. However, there is a great variation in color combination for the various parts. A comparison of pleurisy-root with white-leaved milkweed (page 5) and with rough milkweed (page 45) demonstrates a portion of the variability in flower color. The flowers of white-leaved milkweed are creamy-white; in pleurisy-root they are orange except for the yellowish stigmas. *Asclepias tuberosa* occurs in many soil types from Ohio west to Utah and south to Arizona, as well as in the states of the lower Mississippi River valley. The common name is taken from reported use of this plant in treatment of lung ailments.

Flowering time: May to September   Milkweed family

Yellow Milkvetch

Astragalus flavus Nutt.

Yellow milkvetch grows on clay soils of southern Wyoming, western Colorado, northeastern New Mexico, eastern and southern Utah, northern Arizona, and southern Nevada. The plants grow in clumps from 4 to 10 inches tall. Flower color is mostly yellowish-white, as illustrated, but phases with creamy white or pink flowers are known. Individual flowers are about ½ to ¾ inch long. There are three named varieties known within the yellow milkvetch. A variety with pink flowers, var. *argillosus* (Jones) Barneby, is known only from central eastern Utah. The type variety, var. *flavus*, is the most widespread. It is distinguished from the more southern variety, var. *candicans* Gray, by the yellow and slightly larger flowers. All three varieties possess the strong odor of the element selenium, which occurs in the clay to silty soils where these plants grow. In some years when other palatable plants are in short supply, livestock, mainly sheep, will eat these plants and be poisoned by them. Acute poisoning from these and other seleniferous plants will kill the animals in only a few hours.

Flowering time: April to July; September and October
Legume family

47

Sand or Old-man Sagebrush

Artemisia filifolia Torr.

Old-man sagebrush has very narrow threadlike leaves (hence its scientific name). Because of this and its silvery color, it is among the most handsome of all species of sagebrush. Its setting in sandy soil among other warm or hot desert species enhances its charm and soft beauty. Here it is pictured growing with the small-headed matchweed, *Gutierrezia microcephala* Gray. In both, flowering begins in late summer and continues to mid- or late autumn. The flowers of both are arranged in tiny heads. Individually the corollas are quite attractive when viewed under magnification, but the array of flowers and plants and setting is indeed striking. Old-man sagebrush grows in warm to hot deserts in the canyon country; in other places it is a plains or foothills dweller. It occurs from Nebraska and eastern Wyoming southward to Texas and westward to southern Utah, southern Nevada, and Arizona. The range of small-headed matchweed includes southwestern Colorado, southern Utah, southeastern California, Arizona, New Mexico, and western Texas. The sagebrush is eaten to some extent by grazing animals, but the matchweed is usually avoided, possibly because of its offensive flavor.

Flowering time: August to November   Composite family

Big Rabbitbrush

Chrysothamnus nauseosus (Pallas) Britt.

Plants of big rabbitbrush grow from 2 to 6 feet in height or more. The flower heads vary from ¼ to ½ inch in length and individually are not striking. However, the flower heads are borne in clusters and the mass of heads is extremely showy. Flowering begins in midsummer and builds to a yellow or yellow orange climax in September and October. The plants line miles of roadsides and drainages in most of the western United States and southwestern Canada. The stems were used by Indians in the Southwest for making baskets, and extracts of the plant were used to treat a variety of human diseases. The plants are resinous and have a bad odor and flavor, both obvious requisites of a good medicine. The stems contain latex from which rubber can be made; rabbitbrush has been investigated as a commercial source of rubber. *Chrysothamnus*, from the Greek words for *yellow shrub*, is a genus of some 12 species, all of them confined to the western United States. Some phases of the species are palatable to livestock, but generally the plants are avoided. Big rabbitbrush occurs in many varieties and forms, in practically all soil types from British Columbia to Saskatchewan and south to Texas, New Mexico, Arizona, California, and northern Mexico.

Flowering time: July to November   Composite family

Palmer Cleomella

Cleomella palmerana Jones

The cleomellas are annual plants which occur in the western United States and northern Mexico. About a dozen species are known. In many features the Palmer cleomella resembles the yellow bee-plant, but the plants are smaller (rarely more than a foot tall) and have short, few-seeded pods. The flowers resemble those of the mustard family. However, the stamen number in members of the caper family is commonly eight (not six) and the fruit is not divided by a thin membrane into two cavities as it is in members of the mustard family. The palmately compound leaves or flower features easily distinguish the genus *Cleomella* from the bee-plants and from species of the mustard family. The plants are common in clay soils. Palmer cleomella occurs in western Colorado and eastern and southeastern Utah. Closely related species are known to occur in Arizona. The species figured here was named by Marcus E. Jones in honor of a benefactor who became his friend after Jones returned a wallet that Palmer had lost on an excursion to Colorado. It is a singular plant that is capable of surviving on saline, gypsiferous, and seleniferous soils.

Flowering time: May to September   Caper family

Yellow Bee-plant

Cleome lutea Hook.

Yellow bee-plant is an annual herb of wide distribution in the western states. It takes advantage of moisture provided by spring rains and begins to flower after a minimum of vegetative growth has occurred. If additional rains are not forthcoming, as is frequently the case, then only a few flowers and fruits are formed. The easily distinguished folded seeds lie dormant until sufficient moisture is available. However, if more rainfall occurs or the moisture supply is abundant, the plants continue to grow and flowering continues through most of the summer. Thus, the size of the plants varies from a few inches to 2 to 3 feet in height and width. Because of its long exserted stamens, yellow flowers, and slender pods, the yellow bee-plant resembles the prince's-plume; but the stamen number, palmate leaves, and fruit a unilocular capsule prove to be differential features. The plants grow in many soil types, ordinarily at lower elevations in desert shrub, pinyon-juniper, and mountain brush vegetative types. A second species, *Cleome serrulata* Pursh, occurs widely in the canyon country. It is easily distinguished by its rose purple flowers and larger petals. Both plants were eaten as greens, both raw and cooked, by the Indians.

Flowering time: April to September   Caper family

Yellow Cryptantha

Cryptantha flava (A. Nels.) Payson

Yellow cryptantha is the showiest species of cryptantha in the Colorado drainage system. Mostly the plants occur in sandy soils where the bright yellow flower stalks stand in bold relief against a drab background. The plants grow from 6 to 15 inches tall and often form clumps about as wide. The plants are armed with sharp hairs which can penetrate the skin and cause minor irritation. The limb of the flower is from about ¼ to ⅓ inch wide. When first opening, the limb is white in color, but by the second day the color has changed to yellow. Fruit consists of four one-seeded nutlets which break apart at maturity. Each nutlet is highly ornamented, and the features of the nutlets have been used as primary diagnostic features between species within this genus. The habitat is mainly in the desert shrub and juniper-pinyon vegetative types in Wyoming, Colorado, Utah, Arizona, and New Mexico. Soils inhabited by yellow cryptantha are mainly sandy, but it occurs occasionally in silty and clay soils also. This species is common in rimrock situations in the canyon country.

Flowering time: May and June   Borage family

Newberry Corkwing

Cymopterus newberryi (Wats.) Jones

Among the earliest of the flowering plants in the canyon country are the members of the genus *Cymopterus*. These members of the parsley or carrot family belong to a group of plants known as prevernal. They flower in later winter and early spring, with the inflorescence expanding as the leaves unfold on the surface of the ground. The generic name *Cymopterus* is taken from the Greek words for *wave* and *wing*, an indication of the undulate margin of winged fruits in some species. It is a genus of about 40 species, all of them confined to western North America. The plants grow from perennial tap roots, which are often thickly fleshy, like a carrot or a sweet potato. The tuberous roots of some species have been used for food by Indians and by others also. The flavor is mild and the texture is good, but the roots are difficult to harvest. Often the species grow in gravelly or clay soils whose concretelike texture makes digging very difficult. Their use as food, except in emergencies, is hardly justified. Newberry corkwing is listed among the threatened plants of the United States, but it is widespread and common in southeastern Utah and northern Arizona. Clusters of tiny flowers are commonly only 1 to 2 inches broad.

Flowering time: March to June   Parsley family

Green Mormon-tea

Ephedra viridis Coville

This plant is known by many common names; among them are Brigham-tea, drover's-tea, and Mexican-tea. All of these names call attention to the use of this plant in making a refreshing hot drink. The green stems of the plant, whether fresh or dried, are used in making the tea. Asiatic species of the genus *Ephedra* have long been used as the source of the drug ephedrine, and American plants possibly contain minute amounts of this alkaloid. It functions as a vasoconstrictor (i.e., it causes blood vessels to shrink) and relaxes smooth and cardiac muscle action. Because of this, the drug has been used as a treatment in postchildbirth recovery. Tea made from green Mormon-tea is an attractive yellow orange. Its flavor is enhanced by adding cream and sugar, and the tea does seem to reduce symptoms of mild colds and allergies. The plant is not a flowering one; rather it is a cone-bearing plant which is more closely related to pines and other gymnosperms. The plants, which resemble the horsetails (*Equisetum*), are low to mid-sized shrubs. Individual plants are either male or female (the one figured here is male). Green Mormon-tea occurs from southern Colorado west through southern Utah to Nevada and south to Arizona and southern California.

Cones maturing: May and June   Joint-fir family

Bottle-plant

Eriogonum inflatum Torr. & Frem.

Bottle-plant is one of the botanical curiosities of the arid lands of the Southwest. The plants are annuals, or in some places they are perennials. Annuals take advantage of the moistures in years when precipitation is abundant and clothe the landscape with a verdure that is seemingly out of place in this land of soil and stone unopposed by vegetation. Bottle-plant is one of the opportunists, and in those unusual years the seeds germinate in autumn or early springtime. A basal cluster of leaves is formed which is appressed to the soil surface, and then, as the warmth of springtime offers hope of summer, the stem elongates rapidly and the branches of the inflorescence develop. If moisture is sufficient, the stems become quite robust and several levels of branching occur. This results in a plant with a great umbrellalike summit on which are displayed numerous tiny yellow flowers. The stem below each major branching joint expands to form a swollen, bottlelike hollow structure. These impressive swellings call attention to the plant, and they persist as dried skeletons for several years. Even in very dry years, bottle-plant can be found growing in some portions of the region. The plants are 1 to 3 feet tall, grow in clay, gravel, and sandy soils, and occur from Colorado and Utah south to Arizona and Baja California.

Flowering time: March to October   Buckwheat family

Wallace's Woolly-daisy

Eriophyllum wallacei Gray

In the warm to hot desert lands of southwestern Utah, western Arizona, southern Nevada, and southern California, there occur a series of diminutive annual species of woolly-daisy. They grow in sandy or gravelly soils in the creosote brush vegetative type. Often a second species with white ray flowers, *Eriophyllum lanosum* Gray, grows with Wallace's woolly-daisy. That plant, whose name can be translated as "the woolly woolly-leaf," is distinguished on some minor technical features as well as by the white rays. The genus contains about a dozen species, and all of them are confined to the western United States. Plants of the Wallace's and the woolly-daisies each grow from about ½ to 4 inches tall or rarely more. They can have a single stem, or many stems by branching from near the base. The size and number of plants are controlled by the amount of moisture present in early spring. In moist years the bright flowers of Wallace's woolly-daisy carpet the areas between desert shrubs with hues of yellow. When moisture is sufficient to induce germination of seeds, but insufficient to allow for much growth, then only a tiny single flower head will be produced. This often dwarfs the minute plant on which it is produced. Flower heads of both species discussed here vary in size from about ¼ to ½ inch in width.

Flowering time: February to June   Composite family

54

Western Wallflower

Erysimum asperum (Nutt.) DC.

Wallflowers grow throughout most of the canyon country and are widespread elsewhere in western North America. Of the several species recognized in the region, the most handsome is the western wallflower. It begins to flower in early springtime, when most other flowers are still in bud, and tends to add a touch of bright yellow to the drab landscape. The flowers have four sepals, four petals, six stamens, and an ovary with two cavities. Because the appearance of the flowers in face view resembles a cross, the family is known by the scientific name Cruciferae. Ordinarily the petals are yellow, but populations with orange to burnt orange and maroon are known. The plants grow from 4 to 30 inches tall and flowers are from ¼ to ½ inch wide or rarely larger. The pods which develop from the ovaries are elongate, slender structures up to 3 inches long. The stems and leaves bear peculiar Y-shaped or pick-shaped hairs which are attached in the middle. The plants occur in numerous habitats in most of the vegetative types in the canyon lands.

Flowering time: March to September   Mustard family

Anomalous Sunflower

Helianthus anomalus Blake

The sand dunes and sandy drainages in west-central, southeastern, and southern Utah and in northern Arizona are sparingly clothed with plants that are uncommon or lacking in other southwestern states. Plants which have such a limited distribution are known as endemics. The anomalous sunflower is such a plant. It was overlooked by botanists until the 1930s, when it was finally named and described on the basis of plants collected in the vicinity of the Henry Mountains of Utah. The plants are easily distinguished from all other species of western sunflower by the pale greenish yellow foliage, narrow leaves, and elongate, narrow bracts around the flower head. The plants are annual and their height depends on moisture abundance. Flowering plants only a few inches tall are known, but when moisture is plentiful they will grow to 4 feet tall or more. The flower heads are from 1 to 3 inches wide. The genus *Helianthus*, from the Greek words for *sun* and *flower*, contains more than 60 species. All are native to the New World, but they are now widely distributed as weeds or as cultivated plants in other portions of the world.

Flowering time: June to September   Composite family

Kaibab Bitterweed

Hymenoxys subintegra Cockerell

Hymenoxys is a genus of about 20 species of the western United States, and some additional ones in South America. Only a few species are of broad distribution. The remainder of the species are more or less restricted. The Kaibab bitterweed belongs to this latter group. It is known only from northern Arizona, entirely within Coconino County, where it grows in dry soils in coniferous woodlands. Such related species as *Hymenoxys acaulis* (Pursh) Parker and *Hymenoxys richardsonii* (Hook.) Cockerell are broadly distributed in the western states, where they grow in many habitats. Commonly *Hymenoxys acaulis* occurs in the juniper-pinyon woodlands and in mountain brush vegetative types, while *Hymenoxys richardsonii* is more common in sagebrush and woodland types of higher elevations. Some of the species are known to be poisonous to livestock. Kaibab bitterweed grows from 8 to 15 inches tall and has flower heads from ¾ to 1½ inches wide. It is easily separated from closely related species by the entire or merely 3-lobed leaves which are silvery in color because of a covering of hair.

Flowering time: June to September   Composite family

Long-bracted Deer-vetch

Lotus longebracteatus Rydb.

The genus *Lotus* consists of about 150 species of very broad distribution, mainly in the north temperate region. Some of the Old World species are cultivated as forage plants. Native southwestern species are adapted to survival in a region where lack of water through most of each year is controlling. Some are annual and others perennial in duration. Still others, such as the long-bracted deer-vetch, flower the first year as if they were annuals but also continue to grow and blossom for one or more additional seasons. They are adapted to survive through the dry portions of the year by retreating into a physiological dormancy. Long-bracted deer-vetch is a local endemic of southwestern Utah and north-western Arizona. It has been regarded as a hybrid between two other species, but this does not seem to be the case. Flowers are about ½ an inch long and are typical of members of the legume or pea family. The banner is bicolored, yellow on the front and orange red on the back. Pods are slender, resembling very long pea pods. They open on drying and the seeds fall free. Plants are eaten by grazing animals, although they are seldom present in such abundance to make a large contribution to total diet. The stems are commonly 4 to 10 inches long, and at length they recline on the soil surface.

Flowering time: March to May   Legume family

Fremont Mahonia

Mahonia fremontii (Torr.) Fedde

Also known as Fremont barberry or holly-grape, this is a beautifully attractive flowering shrub of the sand-rock lands of the Southwest. It was named to honor the sometime botanist, ambitious explorer, and presidential aspirant John Charles Fremont. The leaves are evergreen, pinnately compound, leathery, and armed with marginal spines. In winter the leaves take on a purplish cast and the dull purplish shrubs mark that season of dullness in the vegetation. New leaves form on the twigs in spring and are displayed along with the bright clusters of flowers. Floral parts are arranged in three's; there are six each of sepals, petals, and stamens. The anthers open by means of terminal pores, a feature found in few other families of flowering plants. The fruit of Fremont mahonia is a few-seeded, hollow, semi-dry berry. The thin, fleshy fruit coat is tart and of excellent flavor. In seasons when fruit is abundant, they are eaten in large quantities by both coyotes and fox. This accounts for purple-stained fecal pellets of these animals far distant from the plants that produced the fruit. Seeds removed from such pellets germinate readily. Wood of Fremont mahonia is bright yellow in color and very attractive. It checks badly on drying, and its use is limited. Distribution of the species is from Colorado and Utah south to New Mexico and Arizona.

Flowering time: April to June   Barberry family

58

Small Blazing-star

Mentzelia multiflora (Nutt.) Gray

There are several species of blazing-star in the western states. Some, such as the smooth-stemmed blazing-star, *Mentzelia laevicaulis* (Dougl.) T. & T., are very showy. Some others, as *Mentzelia albicaulis* Dougl., have very tiny, almost inconspicuous flowers. Those of the small blazing-star are intermediate in size but exhibit a charming splash of yellow in the desert scene. The plants commence flowering when they are only a few inches high and often continue to produce flowers long after the leaves have withered and the senescent plant is covered with mature and dehiscing capsules. The leaves are pubescent with minutely barbed hairs, causing them to adhere tightly to clothing or animal hair. Because of this the plants often go by the name of stick-leaf. The plants are either biennial or very short-lived perennial plants. Following flowering, the petals and stamens wither and finally fall away. The sepals are persistent at the apex of the inferior ovary. Alternating with the five bright petals are five slender, petallike stamens. Small blazing-star is distributed from Wyoming and Utah southward to Arizona, California, and Mexico.

Flowering time: February to October   Loasa family

Desert Evening-primrose

Oenothera primiveris Gray

The name *Oenothera* was first used by the Greek philosopher Theophrastus, the student of Aristotle and author of the work "On the history of plants." That work appeared sometime between 370 and 288 B.C., and the plant to which the name belonged is lost to history. Now it is the scientific name of a genus of almost 100 species of North and South American plants. The large-flowered species are among the most beautiful of our southwestern plants. Yellow or white to cream are the predominant colors of the flowers in our region (see pages 13 and 15). The desert evening-primrose is a winter annual. Its seeds germinate following autumn rains and form a small rosette of leaves which overwinters. Sometimes, however, the seeds begin growth following early spring rains. The size of the plant and the flower number are dependent on the amount of moisture. Tiny plants often produce only a single flower which is larger than the rest of the plant, whereas plants which are hardier bear numerous flowers. The flowers vary in size from 1 to 2½ inches wide. Desert evening-primrose inhabits sandy soils in hot deserts in southern Utah, southern Nevada, California, Arizona, New Mexico, and Texas.

Flowering time: March to May   Evening-primrose family

Many-spined Prickly-pear

Opuntia polyacantha Haw

The flowers of the prickly-pears are among the most beautiful of flowers. They occur in shades of pink, orange, and yellow within the same species, and all have refractive cells in the petal surfaces which tend to make them difficult to photograph. The human eye is more sensitive than most combinations of camera and film, and by comparison to actual appearance, photographs of cactus blossoms are likely to appear second best. As in other cacti, the job of manufacturing food is left to the fleshy, green, padlike stems. Leaves are borne on young, developing stems. However, the leaves are tiny, fleshy, hornlike structures which soon fall away. The fruits of this prickly-pear are dry at maturity and hence are not used as human food, although the pads have been eaten in times of emergency. The spines must either be burned or peeled away. The pulpy interior portion of the pad is mucilaginous and astringent. The flowers, which remain fresh for only one day, are 2 to 3 inches wide. The stamens in the opuntias are irritable, closing inward when touched at the base. The plants grow in many habitats from the desert shrublands upward to the sagebrush vegetative type in much of the western United States.

Flowering time: April through July   Cactus family

Newberry Twinpod

Physaria newberryi Gray

The genus *Physaria*, represented by some fourteen species, is restricted to western North America. The name is from a Greek word for bellows, a reference to the inflated balloonlike valves of the fruiting pod (a silicle). The Newberry twinpod is characteristic of most of the species in being a low-growing plant which forms clumps by branching from the root crown. The herbage bears intricate, beautiful branched hairs which give a silvery cast to the plant. The individual flowers are about ¼ to ½ inch wide and are borne in showy clusters. The fruit begins to mature as the flowers fade and ultimately develops into an inflated, two-compartmented structure resembling a pair of tiny balloons joined along one side — hence the name "twinpod." The pods demonstrate one phase of variation in fruit type in the mustard family. Fruits vary greatly within the family and are a fundamental feature in distinguishing the genera within the mustard family. Newberry twinpod is found in many soil types in Utah, New Mexico, Arizona, and Nevada.

Flowering time: April to June   Mustard family

Many-lobed Groundsel

Senecio multilobatus T. & G.

Senecio is one of the largest genera of flowering plants, consisting of more than 1000 described species. It is also one of the most broadly distributed genera in the world. The genus is represented in the canyon country by more than two dozen species. All have bright yellow to orange or yellow orange flowers in heads surrounded by a single row of greenish bracts. The individual flowers are small and by themselves inconspicuous, but the clusters of heads are quite showy. All species are eaten by livestock and by wildlife to some extent, but they are not considered to be of great importance as range plants due to their small proportion in the total vegetation. Many-lobed groundsel grows in sandy, clay, or gravelly soils in the pinyon-juniper, sagebrush, mountain brush, and forests at low to moderate elevations in the mountains. The area of distribution is from Wyoming, Utah, and Nevada southward to Arizona and New Mexico. Stems are from 6 to 15 inches tall, and flower heads are ½ to ¾ inch wide. The generic name *Senecio* is derived from a Latin word for *old man*, a reference to the whitish or grayish pubescence of some species.

Flowering time: April to August   Composite family

Silver or Round-leaved Buffaloberry

Shepherdia rotundifolia Parry

The silver buffaloberry is ordinarily a rounded shrub 2 to 5 feet tall and equally as wide or wider, but where they grow in alcoves or along steep canyon walls they sometimes become sprawling shrubs that conform, cascadelike, with the angle of the downhill slope. The plants grow on mesas and canyon slopes in the pinyon-juniper woodlands. They are also common as components of the hanging garden vegetation in canyons at elevations below the pinyon-juniper zone. The shrubs are compact, and the silvery leaves are arranged so as to appear almost like armor. Both leaves and twigs bear delicate branched hairs that resemble tiny parasols with radiating arms. Since the flowering takes place early in the spring, most people do not see the clusters of bright flowers which resemble those of the related Russian olive (*Elaeagnus angustifolia* L.). The fruits are seldom seen either. They resemble the fruits of the Russian olive in size and are silvery in color because of the presence of numerous hairs on the surface. Silver buffaloberry grows in southern Utah and northern Arizona. It is beautifully displayed along the switchbacks below the tunnel in Zion Canyon.

Flowering time: March to May   Oleaster family

Small-leaved Globe-mallow

Sphaeralcea parvifolia A. Nels.

The species of globe-mallow are in the same family as hollyhock and hibiscus, and their flowers are similar in structure. For the most part, the globe-mallows are easily recognized by their salmon- or orange-colored petals. Occasional specimens have white or even pinkish flowers, but they are exceptions. Species with orange- or salmon-colored flowers are rare in the Southwest; thus, the species of *Sphaeralcea* are exceptional. The individual flowers of the small-leaved globe-mallow are only about ½ inch wide, but they are borne in clusters along the upper part of the plant and are very showy. The powder-puff-like cluster of stamens in the flower center adds to the color and beauty of the symmetrical blossoms. The stamens are numerous and are united by their filaments into a sheath which encloses the styles. This combination of characteristics distinguishes members of the mallow family from all other plant families. The plants grow from 1 to 2 feet tall and frequently have several stems from the root crown. They occur in a variety of soil types in western Colorado and Utah to New Mexico, Arizona, and eastern California.

Flowering time: April to October   Mallow family

Prince's-plume

Stanleya pinnata (Pursh) Britt.

Prince's-plume is one of the most strikingly beautiful of canyon country wildflowers. The plants are from 2 to 4 feet tall, having flower clusters up to 2 feet long with up to 1 foot's length displaying open flowers. The structure of the individual flowers adds to the beauty. The stamens protrude from the flowers, giving a delicate, almost lacy appearance, and later the long-stalked pods protrude in bottlebrush fashion. This plant is, however, a Pandora's box, because it is poisonous to livestock. If other palatable vegetation is present, grazing animals avoid prince's-plume. The poisonous agent in *Stanleya* is selenium, which is concentrated in the plant from the soil. *Stanleya* is known to grow only on soils containing selenium and is thus designated as a primary indicator of that poisonous element. Despite the poisonous nature of prince's-plume, Indians in the Southwest utilized it as a potherb, pouring off the boiling water and renewing it a few times before eating the cabbagelike, cooked remains. Plants are most common on soils derived from shales, midstones, and siltstones. They are widely distributed from North Dakota to Oregon and southward to California, Arizona, New Mexico, and Texas.

Flowering time: April to September   Mustard family

Long-spined Horse-brush

Tetradymia axillaris A. Nels.

Growing in sandy or clay soils in the hot deserts of
southwestern Utah, southern Nevada, western Arizona, and
southern California, the long-spined horse-brush is a rounded
to hemispheric bush up to two feet tall or rarely more. The
white hairy stems and long spines formed from modified
leaves are distinctive features of this species. Foliage leaves
are formed on short axillary branches. Of the six to ten
species known in this exclusively North American and Mexican
genus, only two are common in the canyon country. The
widespread species *Tetradymia canescens* DC. is present at
middle elevations in the region. It can be told at once from
the long-spined horse-brush, which occurs at much lower
elevations, by its 4-bracted, 4-flowered heads (from which the
scientific name is derived because they are tetramerous).
Those of *Tetradymia axillaris* regularly bear more of both of
these structures. Additionally, the flower heads of the long-
spined horse-brush are solitary in the axils of the leaves
instead of in clusters at the ends of the branches. At least
some species of horse-brush are poisonous to livestock,
especially to sheep, in which they produce a fatal affliction
known as "big head."

Flowering time: April to June   Composite family

Naked Greenthread

Thelesperma subnudum Gray

The genus *Thelesperma* is represented in the Four Corners
region by only two or three species, but the handsome plants
of naked greenthread more than make up for the low number
of species. The generic name is from the Greek words for
nipple and *seed*, a reference to the seeds having a nipplelike
projection. The stems, or scapes, of naked greenthread are
commonly 6 to 15 inches tall, and the flower heads are from
1 to 2 inches wide. Disk flowers are small and not especially
showy, but they contrast strikingly with the bright ray flowers
which radiate in cuplike fashion from the periphery of the
disk. The stems are often covered with a waxy bloom which
gives the appearance of succulence to the stems and the
united inner bracts that surround the flower bases. The plants
grow in a variety of soil types, but they are common to
abundant on clay soils associated with shale formations in
southern and western Colorado, New Mexico, Arizona, and
eastern and southern Utah. Seeds lack the umbrella type of
pappus characteristic of many composites, and these are
distributed by being shaken from the head by wind.

Flowering time: May to September   Composite family

Yellow Salsify

Tragopogon dubius Scop.

Yellow salsify is much like the purple-flowered salsify or goatsbeard, *Tragopogon porrifolius* (page 82) in size and habit. Both belong to that section of the composite family whose members possess milky juice. The dandelion also belongs to this section of the family. All members are similar in having only a single flower type in the head of flowers; all are strap-shaped and bilaterally symmetrical. The flowers open in early morning and close by about midday of the day following. The globose fruiting heads possess an intricate beauty, and the parachute mechanism allows for efficient dispersal of seeds. This accounts for the wide distribution of the species in North America and in the Old World, from which it was introduced. The parachute consists of plumose pappus hairs which spread at right angles from the summit of the long-beaked achene, or fruit, of this plant. The pappus hairs stand erect around the base of the individual corollas in flower but spread to the parachute form upon maturity of the fruits. Plants grow in disturbed sites, especially in agricultural regions, but they are widely scattered in nonagricultural areas as well.

Flowering time: May to September   Composite family

Rough Mules-ears

Wyethia scabra Hook.

Named in honor of the nineteenth-century western explorer Nathanial Wyeth, the genus *Wyethia* consists of some 14 species which are restricted to western North America. Most species are foothills or mountain plants; only rough mules-ears inhabits the sandy and arid canyon country. In the Upper Colorado River Basin there are numerous plants which grow only in sand dunes or in other very sandy sites. Among them are the rough mules-ears. Similar to other plants of dune areas, rough mules-ears have the capacity of adjusting to differences in soil level, particularly when they are covered by an advancing dune. The plants root along the stems, and new shoots grow above the surface. The plants grow commonly to a height of 1 to 2½ feet tall and occur in clumps to several feet wide. The flower heads, which resemble those of sunflowers, are from 2 to 4 inches in width. The leaves are rough-hairy, and it is from the nature of the leaves that both the scientific and common names have been derived. Rough mules-ears occurs in eastern and southeastern Utah, western Colorado, and northern Arizona.

Flowering time: June to October   Composite family

The small-flowered
columbine (*Aquilegia
micrantha* Eastw.) is an
inhabitant of the hanging
garden flora of south-
western Colorado,
southeastern Utah, and
northeastern Arizona. In
well-developed hanging
gardens this columbine
grows along the mound of
detritus at the foot of the
moist wall of the alcove.
The columbine grows
from 1 to 2 feet tall, and
the flowers range in size
from about ½ to 2 inches
in length. The spurred
petals are unique to this
radially symmetrical
flower. At maturity each
of the fruits (follicles) splits
along one side and
releases numerous
tiny seeds.

Colorado Canyon Milkvetch

Astragalus kentrophyta Gray

In the Old World there are many representatives in
Astragalus and other legume genera, such as *Caragana*, in
which the leaves are reduced to spines. None of the
American species has followed this exact pattern, but the
milkvetch figured here is armed nevertheless. The leaflets are
thick and leathery, and each of them is equipped with an
apical spine. They present a formidable task to all that would
try to eat or to collect them. The tiny flowers give way to very
short but typical legumes. There are several named varieties
known within the species. For the most part the varieties are
low growing or prostrate in habit, but the var. *elatus* Wats.
produces erect stems which simulate some phases of the
Russian thistle, *Salsola pestifer* A. Nels. That variety grows at
middle to moderate elevations, as in the vicinity of Bryce
Canyon National Park. The low-growing var. *implexus* (Canby)
Barneby occurs at very high elevations, often on calcium-rich
soils. The variety featured here, var. *coloradoensis* Barneby, is
a plant of low elevations; it was described from plants
collected at Lee's Ferry, Coconino County, Arizona, by Marcus
E. Jones in 1890. Its distribution is from northern Arizona
northward along the canyons of the Colorado to Emery
County in central eastern Utah.

Flowering time: April to June   Legume family

Wild-cabbage

Caulanthus crassicaulis (Torr.) Wats.

In *Caulanthus crassicaulis*, the flowers appear to rise directly
from the stem. Because of this the generic name *Caulanthus*
(Greek for *stem* and *flower*) was derived. The genus is a
small one, with only about a dozen species, most of them in
California. The name *crassicaulis* is likewise descriptive of the
thickened stem that often develops in this species. The stem
becomes expanded in balloonlike fashion similar to that of
the bottle-plant (page 52). The swollen stem is expanded
over a larger portion, often affecting the lower part of the
inflorescence as well. The fruits are siliques, which are very
slender and 4 to 6 inches long. These are borne erect or
ascending, and in mature plants they give a broomlike
appearance. Plants are biennial or perennial, and often the
dried, straw-colored stems of the previous year still adorn the
plants when they flower in the springtime of the current year.
Heights of up to 2 feet are not uncommon, but flowering
stems as low as 8 to 10 inches are frequent. Flowers are
about ½ an inch long. The plants are distributed in low
elevations from Oregon and Idaho south to Arizona and
California. The plants were eaten by Indians as potherbs and
also fresh.

Flowering time: April to June   Mustard family

72

Chickory; Succory

Cichorium intybus L.

Scientific names are considered to be Latin regardless of origin of the names. This is the case with the name *Cichorium*. It is Latin although based on the Greek name *kichore*, which is probably from the old Arabic word *chikourgeh*. The genus contains about eight or nine species, all of Old World origin. The stems of chickory grow to a height of 2 to 3 feet or more, and by profuse branching they form clumps to 2 feet wide. The flowers open in early morning and close before midday. Thus, the plants are enjoyed most by those who arise before noon. With flowers closed, the plants take on a ragged appearance with the zigzag skeletonlike stems and small bracts and foliage leaves. The flower heads range in size from ¾ to 1½ inches wide or more. Chickory was introduced from Europe for its roots, which are used as a coffee substitute, and for its leaves, which are used in salads. A closely related species, *Cichorium endivia* L., is the endive of commerce. Chickory has escaped from cultivation and persists along roadsides, chiefly in agricultural regions. It is widely distributed in North America.

Flo·vering time: June to September   Composite family

Columbia Virgin's Bower

Clematis columbiana (Nutt.) T. & G.

This species and the closely related and more common canyon lands entity, *Clematis pseudoalpina* (Kuntze) A. Nels, form a distinctive pair of native *Clematis* species. The flowers lack petals, but the sepals simulate petals both in texture and in color. The trailing vines and bright blue to violet sepals, which are 1 to 2 inches long, combine to make this a very striking plant. The fruits, which develop from the numerous separate ovaries, form a globose silken tuft to 4 inches long by elongation of the hairy styles. Both species grow on mountain slopes or canyon sides, often in deep shade where they clamber over rocks and other vegetation. Neither of them forms the robust growth of the species with white to cream flowers which is found in the canyons at low elevations. That species, *Clematis ligusticifolia* Nutt., often festoons shrubs along drainages to 10 feet in height, and the silky fruiting clusters give a silvery tapestry effect to late summer and autumn scenery. *Clematis columbiana* occurs from British Columbia and Alberta south to Utah and Colorado; *Clematis pseudoalpina* from Montana and South Dakota south to Arizona and New Mexico; and *Clematis ligusticifolia* from western Canada to North Dakota and south to California, Arizona, and New Mexico.

Flowering time: May to August   Buttercup family

Fleabane Daisy

Erigeron pumilus Nutt.

This common daisy is from 2 to 12 inches tall. The leaves and stems are spreading-hairy, adding to the soft elegance of the plants. The flowers vary in color from white to pink or purple, and the heads are from ¾ to 1½ inches in diameter. Fleabane daisy occurs in desert shrublands, grasslands, pinyon-juniper woods, and upward to the mountain brush zone of the foothills. This species is known from Idaho and Saskatchewan southward to California, Arizona, New Mexico, and Kansas. There are several morphological variants known which have received names. This is not surprising in a plant with such a broad horizontal and vertical area of distribution. The genus to which it belongs, *Erigeron*, is a moderate to large one, with perhaps 200 species. It is recognized to be a taxonomically difficult genus. This is partially due to the large size of the genus and partly due to the fact that many of the species do not reproduce in a regular manner. As in pussytoes (page 4) many of the species of *Erigeron* set seed without pollination having taken place, and this leads to consequences which make the interpretation of species very difficult.

Flowering time: April to July　　Composite family

Heronsbill; Filaree

Erodium cicutarium (L.) L'Her.

Introduced into western North America from the Mediterranean region of the Old World sometime in the nineteenth century, heronsbill spread quickly through the low-elevation arid lands of the Southwest. This plant (sometimes called "storksbill") is ordinarily a diminutive plant with leaves and stems 2 to 5 inches long, but under conditions where moisture is abundant the prostrate stems reach a length of 2 to 3 feet. The flowers are only ¼ to ½ inch broad, but their attractive bright pink color allows the plant to be seen easily. Elongating at maturity, the ovary produces the distinctive beaklike fruit characteristic of the geranium family; the fruit type accounts for both the scientific (Greek origin) and the common name of heronsbill or storksbill. The plants grow in areas where soil and natural vegetation have been disturbed by man, his machines, or his animals. Of the approximately 50 species of *Erodium* known from southern Eurasia, Africa, and Australia, many have been successfully established in other parts of the world. A second species, *Erodium texensis* Gray, is present in the hot deserts of the Southwest, and it would be surprising indeed if other species should not be introduced and become established also.

Flowering time: April to October　　Geranium family

75

Missouri Iris; Blueflag

Iris missouriensis Nutt.

Plants of Missouri iris grow singly or in clumps. In some regions they form solid stands, acres in extent, in moist meadows. The plants grow from 1 to 2 feet high and have two-ranked leaves with parallel veins characteristic of the family. The flowers are from 1½ to 3 inches or more wide; they form an outer whorl of three sepals known as falls and an inner whorl of three petals known as standards. The stamens are three in number, and they are ordinarily covered by the three brightly colored, petallike stigmas. Thus, all floral parts are in threes. Flowers are not so large nor so brilliantly colored as some of their cultivated counterparts, but in their own setting they present a display of subtle charm. Missouri iris is the only species of iris native to the canyon country, where it occurs at middle elevations usually in moist, open sites. The area of distribution is from British Columbia east to North Dakota and south to California, Arizona, and New Mexico. Other species of iris, chiefly *Iris germanica* L., are grown in much of the region. Some have been planted in soil conservation plantings, and others have escaped.

Flowering time: Mid-May to late July   Iris family

Sand-loving Lupine

Lupinus ammophilus Greene

Sand-loving lupine is characteristic of the many species of perennial lupines which are widely distributed in western North America. It grows from 1 to 2 feet in height, and the stems are terminated by flower clusters 2 to 6 inches long. The individual flowers are about ½ an inch in length. The banner (the uppermost petal in legume flowers) has in its center a contrasting patch of color which is called the eyespot. Often the eyespot is bicolored, with yellow and white predominating. Lupines have long been favored as cultivated plants by gardeners, but since some of them are poisonous to livestock, they are regarded with mixed feelings by most stockmen. Sand-loving lupine grows in sandy or gravelly soil in the juniper-pinyon, sagebrush, and mountain brush vegetative types in Utah, Colorado, and New Mexico. The flowers of lupines contain a feature which is different from most representatives of the legume family. Stamens are ordinarily united by their filaments into a bundle of nine, with the solitary tenth stamen lying free along the top of the ovary. In lupines, however, all of the stamens are united into a tube which completely surrounds the ovary.

Flowering time: April to June   Legume family

Dwarf Lupine

Lupinus pusillus Pursh

In the arid southwestern United States there are several species of low-growing annual lupines. Dwarf lupine is characteristic of them. Their abundance in any given locality is determined by the amount of moisture available in the soil when spring arrives. If spring or late winter storms do not provide sufficient moisture, as is often the case, then the seeds fail to germinate and will remain dormant until the rains of succeeding seasons are sufficient to allow for germination. Also, the stature of the plants is, in large part, determined by the amount of water available following germination. If water is in short supply, the plants may produce only a few flowers and even fewer mature seeds; if water is abundant, then the plants respond with luxuriant growth, numerous flowers, and abundant seeds. The plants are commonly only 2 to 8 inches tall. They grow in a variety of soil types ranging from gravel and sand to heavy clays. The species occurs in several phases from Washington to Saskatchewan and southward to Nevada, Arizona, New Mexico, and Kansas.

Flowering time: April to June   Legume family

Dwarf Beardtongue

Penstemon caespitosus Nutt. ex Gray

Penstemon is a genus with a great diversity of form among its many species. Plants with tall and slender stems and with bright interrupted clusters of flowers are the most common types. An extreme phase in the diversity is found in the mat-forming species with flowers formed almost at ground level. Bright-colored flowers borne on a carpet of green to gray green leaves is the characteristic of this group. One is always struck by the apparent anomaly of the penstemon's beautiful flowers borne at ground level. The lower lip of each corolla protrudes and is used as a landing place for the insects, especially certain kinds of bees which serve as pollinators. These insects search for nectar at the base of the corolla tube, where it is secreted by special glands. Dwarf beardtongue grows in mats to 4 feet in diameter, but small plants only a few inches across are common also. Flowers are ½ to ¾ inch long. Plants grow in gravelly or sandy soils in the pinyon-juniper, mountain brush, and sagebrush plant communities, usually at middle elevations. The range of *Penstemon caespitosus* is from Wyoming southward to western Colorado and much of Utah.

Flowering time: June and July   Figwort family

Nelson Phacelia

Phacelia corrugata A. Nels.

Phacelia is a moderately large genus with perhaps 150 species. The number of species is greatest in the western United States and northern Mexico. Some of the species, perhaps most of them, secrete resinous materials from glandular hairs along at least the upper part of the plant. Contact with the glandular secretions has produced dermatitis in some persons similar to that caused by poison ivy. The plants are easily recognizable by their bright, usually blue purple, flowers which are borne in scorpioid cymes (resembling the tail of a scorpion). In these clusters, the flowers are borne in sequence along one side of the axis of the cluster which seems to uncoil from the end. The Nelson phacelia is biennial, or possibly it is sometimes a winter annual. The flowering stems bolt quickly as warmth returns following winter. The first flowers open while nighttime temperatures are still in the freezing range. Soon a mass of flowers, blue purple in color, marks the landscape. Soil requirements are not strict for this handsome plant. While the plants tend to grow on almost any soil type, they are able to grow on some of the soils which other plants tend to find intolerable. They grow especially well on clay soils. Area of distribution is from Colorado and Utah south to Arizona and New Mexico.

Flowering time: March to September   Waterleaf family

Purple-sage

Poliomintha incana (Torr.) Gray

This is the grayish-leaved aromatic plant made famous in the title of a western novel by Zane Gray — *Riders of the Purple Sage*. The plants tend to grow in sand dunes or other unstable sandy soil, where they have the capacity to grow up through the deepening layer of shifting sand. Leaves are opposite and the stems are round to four-angled, as is characteristic of members of the mint family. Flowers are strongly two-lipped, and the two stamens are placed well above the middle of the corolla tube. Despite the attractiveness of the individual flowers, the plants are seldom showy, even when in full flower. The corollas fall within a short time, and usually only a few are open along a stem at any given time. Stems branch repeatedly to form rounded shrubs or irregularly shaped ones to 1 or 2 feet in height. Often much of each plant is buried below the surface of the dune, and only the uppermost branches protrude. The flowers and some portions of the inflorescence have been used in seasoning of food, and there is no reason why the leaves could not be used as in garden sage. Area of distribution includes southern Utah, northern Arizona, southern California, and northern Mexico.

Flowering time: May to September   Mint family

Desert-sage

Salvia dorrii (Kellogg) Abrams

Desert-sage belongs to the largest genus in the mint family. *Salvia* contains perhaps 500 species of almost cosmopolitan distribution, but the main centers of distribution are in tropical and subtropical America. The scientific name is thought to have been derived from the Latin word for *save*, referring to reputed medicinal properties. Certainly the plants are aromatic, and it is from this genus that the garden sage, *Salvia officinalis* L., is harvested for use as a seasoning. Desert-sage is a low, hemispheric shrub with whitish foliage contrasting with purple bracts of the inflorescence and sky blue corollas. The picture formed by this plant, when in full flower, is both striking and pleasing. Adding to the character of this handsome plant is the aroma, which is unlike but similar to that of the garden sage. Rimrock situations at the edge of the pinyon-juniper woodlands, sandy sites in cool desert shrublands, and gravelly soils of hot deserts form the most common habitats for desert-sage. Several named varieties are known within the large amount of variability present within the species. It is distributed from Washington, Oregon, and Idaho to southern Utah, western Arizona, and southern California.

Flowering time: April to June   Mint family

White Horse-nettle

Solanum elaeagnifolium Cav.

The beautiful flowers of this handsome plant belie its
character as a troublesome weed of cultivated land and waste
places in the warmer portions of the Southwest and parts of
the plains states. But this plant is not exceptional in having
beautiful flowers or in being weedy. The genus, with more
than 1000 species, contains several important weeds, some of
them remarkably beautiful. The buffalo-bur, *Solanum
rostratum* Dunal, has striking light yellow corollas about an
inch wide. In both white horse-nettle and buffalo-bur the
stems, leaves, and often other plant parts are armed with
stout, straight prickles or spines which rival rose thorns for
sharpness and ability to penetrate flesh. Stamens with short
filaments and anthers more or less coherent in a conelike
structure around the style in the flower's center are
characteristic of members of the genus. The flattish, open
corolla with rotate limb is also characteristic. The fruit of
white horse-nettle is globose and yellow to black at maturity.
It is a berry, botanically speaking, and resembles a small
yellow or black tomato, to which it is related. White horse-
nettle is distributed from Kansas, Colorado, and southern
Utah south to tropical America, and it is adventive elsewhere.

Flowering time: May to October   Nightshade family

Narrow-leaved Sophora

Sophora stenophylla Gray ex Ives

Sand dunes, sandy benchlands, and sandy drainages in
southern Utah, northern Arizona, and New Mexico are often
inhabited by narrow-leaved sophora. The stems arise from
elongate rhizomes and are capable of surviving in a harsh
habitat where the plants are alternately buried or left dangling
in the air by the shifting sands. Stems develop from buds
along the rhizomes at or near the surface of the sand and grow
from 6 to 18 inches tall or more, but often only the tip of the
branch will protrude above the dune. The brightly colored
flowers and silvery foliage combine to form a striking contrast
with the reddish sand in which they grow. Flowers are $2/3$ to
1 inch long, and vary in color from violet to deep blue purple.
Seeds are borne in elongate slender pods which are more or
less constricted between the seeds. The pods are borne erect
or steeply ascending along the axis of the inflorescence.
Narrow-leaved sophora is the most striking of the herbaceous
species of sophora in the canyon country. The second
species, *Sophora nuttalliana* Turner, is quite inconspicuous by
comparison. Its flowers are much smaller and creamy white
to pale yellowish in color.

Flowering time: April to June   Legume family

Goatsbeard; Oyster-plant; Salsify

Tragopogon porrifolius L.

This purple-flowered species of *Tragopogon* grows to 3 feet tall or more and is a strict biennial. A stout taproot and basal rosette of leaves forms during the first season. The elongated leafy stem grows during the second growing season, flowers, produces fruit, and dies in the autumn of the second year. The flower heads are 1 to 2 inches wide. The roots have been eaten as cooked vegetables, and they are said to resemble oysters in flavor. This resemblance is the source of the common name oyster-plant. The species was introduced from the Old World and has escaped from cultivation. It persists along roadways and margins of cultivated land in much of North America. However, like its relative yellow salsify (page 68), this plant is very efficiently dispersed by the wind-borne umbrella or parachutelike pappus attached to the long-beaked achenes. The genus of the goatsbeard, *Tragopogon* (a direct translation of the Greek words for goat's beard) consists of some 50 species, all of them native to Eurasia and Africa. Only three of these have become established in our region.

Flowering time: May to September   Composite family

Western Spiderwort

Tradescantia occidentalis (Britt.) Smythe

The blossoms of members of the spiderwort family resemble those of the lily family: sepals and petals three each, stamens six, and a three-loculed ovary. The sepals are of different texture and color than the petals in the spiderworts and are commonly of the same color and texture in the lilies. The staminal hairs of this flower and of those of related genera have been used by students of botany to demonstrate streaming of cytoplasm in the purplish cells. The flowers are sensitive to light, closing in the afternoon and opening early the next morning. Plants which were so conspicuous in mid-morning are difficult to find in mid-afternoon. The bright flowers which called attention to them are closed, and the plants blend into the mass of plant species in the sandy soils where this plant grows. Flowers are from about ¾ to 1¼ inches broad and are borne on plants from 6 to 15 inches tall. The region occupied by this species is from Wisconsin west to Montana, Colorado, southern Utah, Arizona, and from Louisiana west through Texas to Mexico.

Flowering time: April to September   Spiderwort family

Kidney-leaved Violet

Viola nephrophylla Greene

This stemless, purple-flowered violet grows to a height of 1 to 5 inches, seldom more. Kidney-leaved violet and other species produce flowers of two types: the showy, open type figured here and an inconspicuous kind which does not open. The flowers of the first type, which are ordinarily cross-pollinated, seldom produce fruit; but the flowers of the second type, which are self-pollinated, practically always produce fruit with viable seeds. Thus, the second flower type (cleistogamous) provides an insurance policy, guaranteeing seed for the following year. The open flowers, which are $\frac{1}{3}$ to $\frac{3}{5}$ of an inch long, provide the possibility of cross-pollination, which is occasionally successful. This allows the introduction of genetic variation which is lessened by self-pollination. For plants in stable habitats it is sometimes an advantage for them to have a genetic stability, but where habitats are changing it is advantageous for the plant to have the capability of genetic variation which will allow the plant to occupy the different habitat type. Thus, violets have compromised in such a manner as to have the best of two worlds. The plants grow in moist, shaded sites and in open meadows, in widely scattered sites over much of North America.

Flowering time: May, June, and July   Violet family

83

Index